# Workshop Technology and Practices

NIPA® GENX ELECTRONIC RESOURCES & SOLUTIONS P. LTD.
New Delhi-110 034

## About the Authors

**Dr. Atul K. Shrivastava,** Dean Faculty, Agricultural Engineering, JNKVV, Jabalpur. He has a doctorate, in Farm Machinery and Power Engineering from IIT, Kharagpur having more than 36 years of experience. He has guided 30 doctorate and 32 master's students and published 191 research papers, 07 books, 03 practical manuals, and 31 bulletins. 5 patents were granted in his account out of them 04 were granted by the Govt. of Germany and 1 from 01 Govt. of India. He is a doctorate fellow of the Indian Society of Agricultural Engineering (ISAE), ICAR, New Delhi, and a fellow of the Institute of Engineers, Kolkata.

**Dr. Subhash Gautam,** Contractual Faculty, Farm Machinery and Power Engineering, College of Agricultural Engineering, JNKVV, Jabalpur. He has a master's in Heat Power Engineering and a bachelor's in Mechanical Engineering. He is the teaching profession for more than 6 years and 01 year of industrial experience. He has published 15 research papers in national and international journals.

# Workshop Technology and Practices

**Atul Kumar Shrivastava**
Dean Faculty
Agricultural Engineering
JNKVV, Jabalpur. Madhya Pradesh

**Subhash Gautam** (M.E. Mechanical)
Farm Machinery and Power Engineering
College of Agricultural Engineering
JNKVV, Jabalpur, Madhya Pradesh

nipa®
NIPA® GENX ELECTRONIC RESOURCES & SOLUTIONS P. LTD.
New Delhi-110 034

NIPA® GENX ELECTRONIC
RESOURCES & SOLUTIONS P. LTD.

101,103, Vikas Surya Plaza, CU Block
L.S.C.Market, Pitam Pura, New Delhi-110 034
Ph : +91 11 27341616, 27341717, 27341718
E-mail:newindiapublishingagency@gmail.com
www: www.nipabooks.com

***For customer assistance, please contact***
Phone: + 91-11-27 34 17 17
Fax: + 91-11- 27 34 16 16



ISBN: 978-93-95319-16-4



Composed and Designed by NIPA®.

# Preface

Welcome to the world of engineering workshop technology and practice. This book is designed to be your comprehensive guide through the fundamental principles, techniques, and practical applications that form the backbone of modern engineering workshops. Whether you are a student embarking on your engineering journey or a professional seeking to enhance your skills, this book is here to provide you with a solid foundation and practical insights into the dynamic field of workshop technology.

Engineering workshops are the heart of innovation and creation in various industries. They are the places where raw materials are transformed into functional products, where ideas are translated into tangible realities. This book aims to bridge the gap between theoretical knowledge and hands-on expertise, offering a balanced blend of theoretical explanations and practical demonstrations. It will empower you to understand the underlying concepts and methodologies that drive the processes within a workshop setting.

**Key Features of the Book**

**Comprehensive Coverage**: This book covers a wide range of topics, from the basics of workshop tools and materials to advanced machining processes and fabrication techniques. The content is organized in a logical sequence, allowing you to build your understanding progressively.

**Practical Approach:** While theory provides the foundation, practical application is the cornerstone of engineering workshops. Throughout the book, you will find real-world examples, case studies, and step-by-step guides that illustrate how concepts are translated into action.

**Visual Aids:** A picture is worth a thousand words. To enhance your comprehension, this book is richly illustrated with diagrams, photographs, and schematics. Visual representations of tools, machines, and processes will provide clarity and facilitate faster learning.

**Problem Solving:** Engineering is about finding solutions to challenges. In line with this ethos, the book includes problem-solving exercises and scenarios that encourage you to apply your knowledge to practical situations. These exercises will sharpen your critical thinking and decision-making skills.

**Safety Emphasis:** The safety of individuals working in engineering workshops is of paramount importance. Safety guidelines, precautions, and best practices are integrated throughout the book to instill a culture of responsible and secure workshop operations.

**Emerging Trends:** Engineering is a constantly evolving field. This book takes cognizance of the latest technological advancements and trends in workshop practices, ensuring that you are prepared to embrace the future of engineering.

# How to Use This Book

This book is designed to be a companion on your learning journey. It can be used as a textbook for formal education, a reference guide for professionals, or a self-study resource. Each chapter builds upon the knowledge gained in previous chapters, so it is advisable to follow the sequence for optimal understanding. However, feel free to explore chapters that align with your specific interests or immediate learning needs.

As you delve into the world of engineering workshop technology and practice, remember that learning is a continuous process. Practical skills are honed through hands-on experience, so make the most of opportunities to apply what you learn in real workshop environments. Embrace challenges, ask questions, and never stop exploring the vast landscape of engineering possibilities that lie ahead.

**Acknowledgements**

Creating a book of this nature involves the collective effort of many individuals. We extend our gratitude to the educators, practitioners, and researchers who have contributed their expertise and insights. Our thanks also go to the readers whose curiosity and enthusiasm drive the pursuit of knowledge and excellence in the field of engineering.

With excitement, we invite you to embark on this journey through engineering workshop technology and practice. May this book empower you to unlock your potential and make meaningful contributions to the ever-evolving world of engineering.

Sincerely

**Authors**

# Contents

# List of Figures

# 1

# Carpentry Shop

## Study of Carpentry Hand Tool

### Introduction

The process of making the wooden components is called carpentry. Carpentry deals with the mechanical work of making wood components such as making beds, roofs, wardrobes, tables, cases, utensils, floors, etc. It manages the structure work, furniture, cabinet making and so on joinery, i.e., preparation of joints is one of the significant activities in all wood works.

The products produced are used in making furniture, musical instruments, and building constructions like doors, windows, desks, chairs, cupboards, etc., of a place utilizing a specific kind of wood with the help of tools known as carpentry tools. Different wooden pieces are connected to make a useful product. Wood is used to make furniture, boxes, doors, windows, industrial floors, stairs, pencils etc.

### 1. Marking & Measuring Tools

These types of tools are designed for accurate marking and measuring because it is essential for carpentry work.

**Figure 1.** Folding rule

#### i) Rules

- **Folding Rule:** The commonly used rules are four-fold box-wood rules ranging from 0 to 60 cm. All four pieces, each length 15 cm are joined together by hinged joints. In this rule, there are graduations marked in millimeters and meters on both sides.
- **Steel Rule:** It is made of steel and is an important tool for linear measurement.

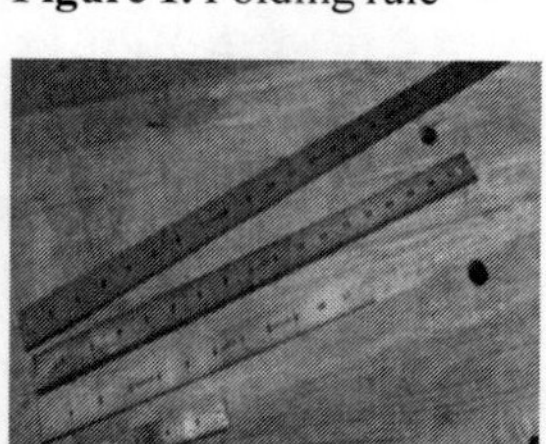

**Figure 2.** Steel rule

i) **Try Square:** It is used for checking squareness, and flatness, measuring dimensions and marking parallel and perpendicular lines over the plane surfaces. It contains a steel blade and a cast iron stock connected at the right angle to each other. It is used to check the perpendicularity of the workpiece. According to the length of the blade, its size varies from 150-300 mm.

**Figure 3.** Try Square

ii) **Bevel Square:** It is used for duplicating, setting, comparing and testing angles. It consists of a wooden or metallic stock fitted with a slotted blade. At any point or angle, we can adjust the blade. The bottom screw is used to tighten the plate position.

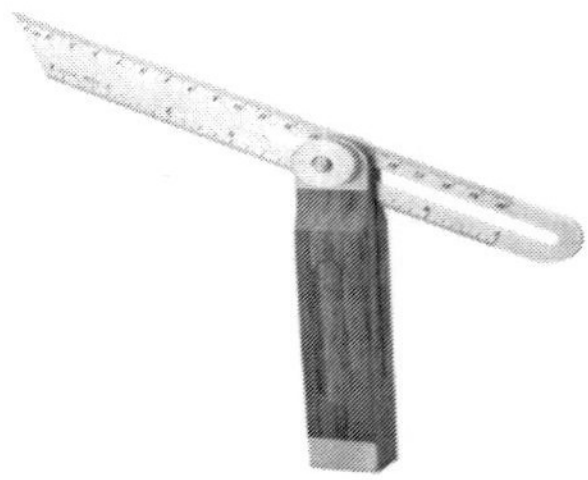

**Figure 4.** Bevel Square

iii) **Mitre Square:** It is used to measure angles of 45 degrees. They are made of all metal with a nickel-plated finish. It consists of a steel blade and rose-wood stock. The blade ranges from 200mm, 250mm, etc., up to a maximum length of 300mm.

**Figure 5.** Mitre Square

iv) **Scriber or Marking Knife:** It is used for marking and locating points and scrubbing lines on surfaces.

- **Gauges:** Used to mark parallel lines to the edge of a piece of wood. This head can slide over stock to set the distance between pins. Screws are used to tighten head over stock.

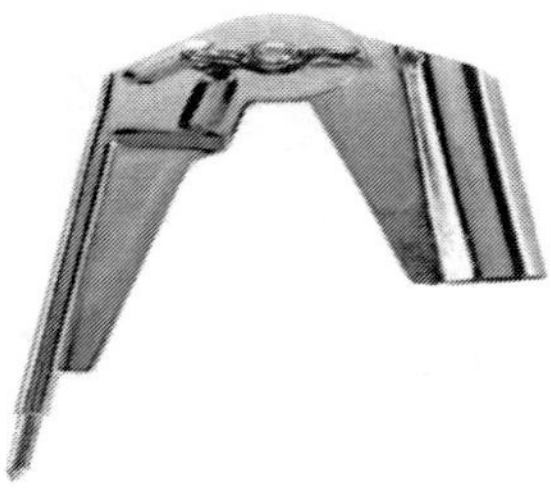

**Figure 6.** Scriber

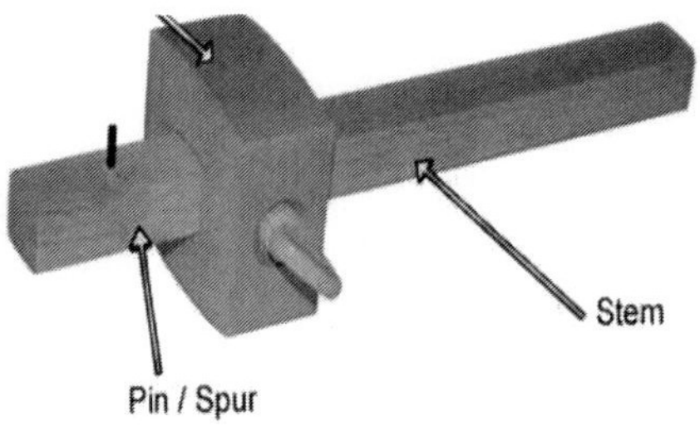


**Figure 7.** Gauges

- **Marking Gauge:** It is made up of wood. It is made up of a wooden stem of a square cross-section and a stock that slides on it that carries a brass liner towards the scrubbing pin fitted in the stem.

- **Mortise Gauge:** It is an improved form of a marking gauge. Uses are marking two parallel gauge lines. And marking mortise and tenon joints.

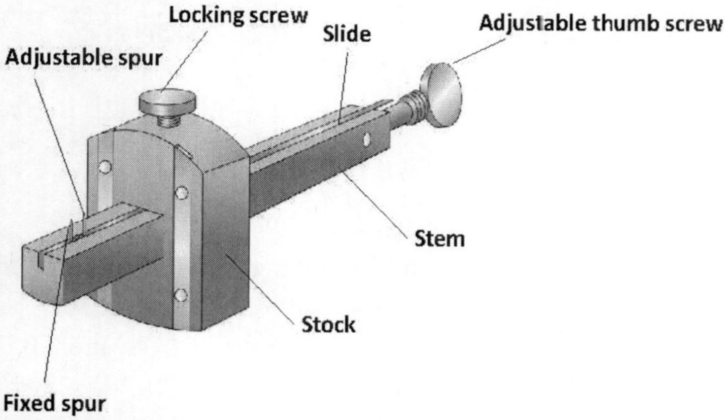


**Figure 8.** Mortise Gauge

vi) **Compass:** The compass has one point sharpened in needle point fashion and the other one is an arrangement for attaching a pencil.

vii) **Trammel:** It is a beam compass with a wooden beam. It is used to take the work that is beyond the scope of the compass.

viii) **Divider:** It is designed like two sharp points like a needle. It is used for transferring dimensions and dividing centers.

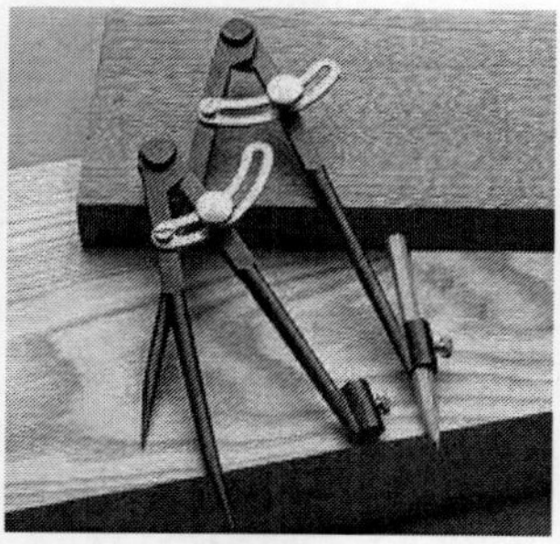

**Figure 9.** Compass

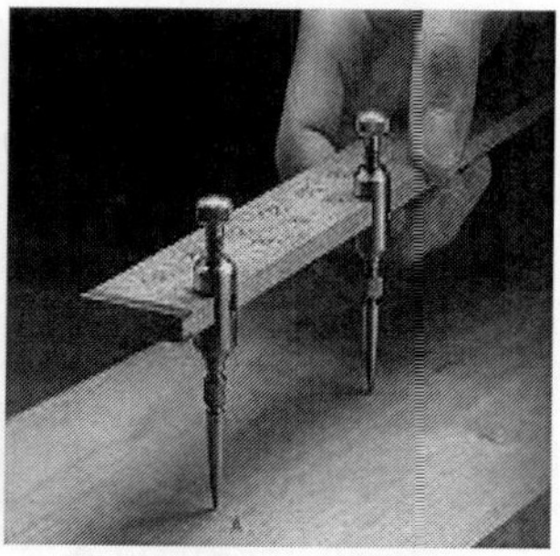

**Figure 10.** Trammel

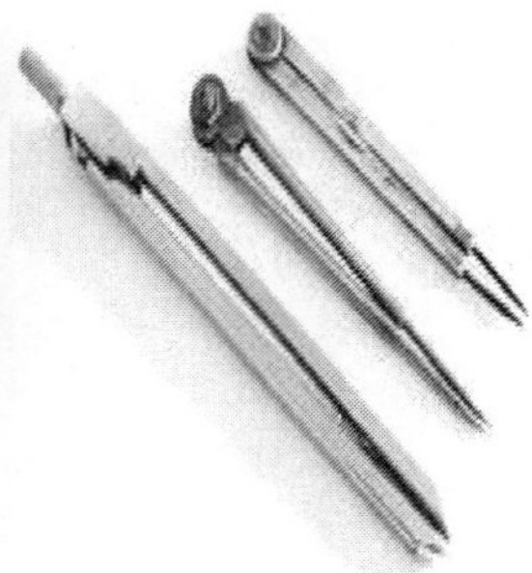

**Figure 11.** Divider

ix) **Caliper:** With the help of this tool, we can measure both the internal and external diameters. These are useful where we cannot measure small lengths.

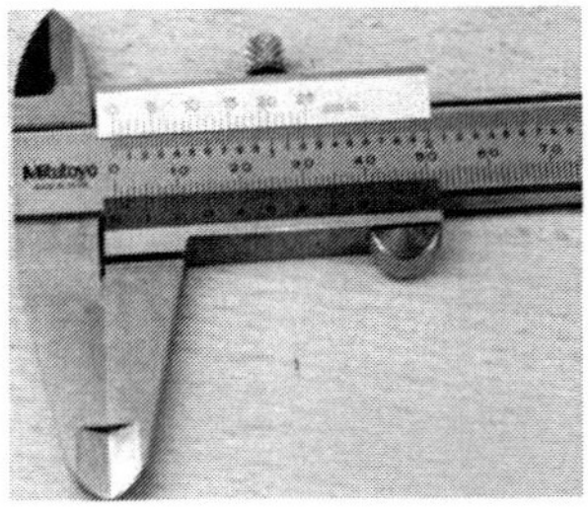

**Figure 12.** Caliper

## 2. Holding Tools

i) **Carpenter's Vice:** It is provided with a jaw that is fixed on a table and it also consists of a movable jaw that can be positioned with the help a screw to move it and handle it. It is used for holding the work for different operations like sawing, planning and chiseling.

ii) **Bar Clamp:** It is made of a steel bar that has the shape of a T-section. It is used for holding frames and tops like wide works.

iii) **C-Clamp:** It is a clamp of C shape and is used for holding small work pieces. These are available in sizes 70 mm to 800 mm. It is used for holding planks.

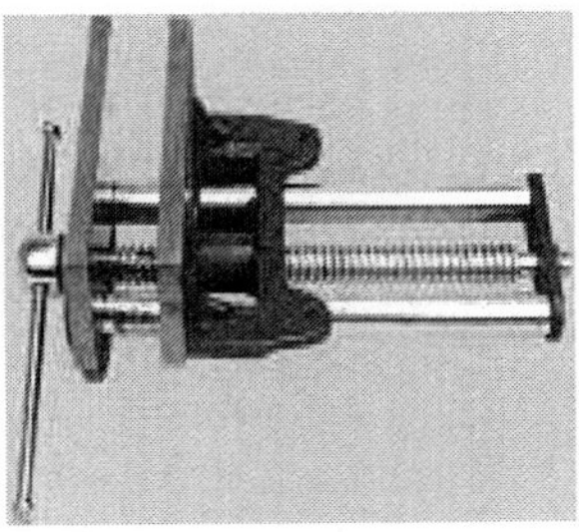

**Figure 13.** Carpenter's vice

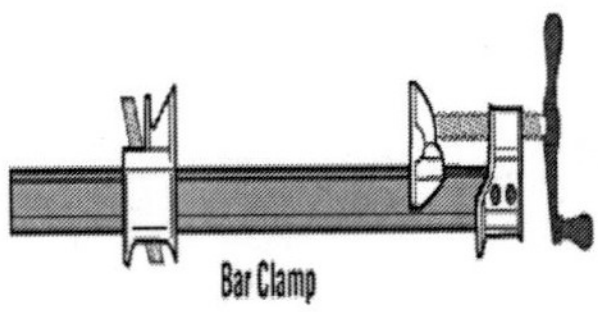


**Figure 14.** Bar Clamp

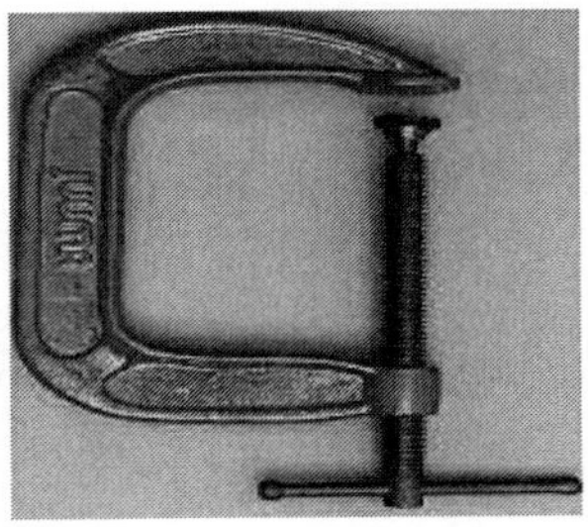

**Figure 15.** C-Clamp

## 3. Cutting Tools

i) **Saws:** It is a cutting tool that is provided with a toothed edge and its working principle is that the toothed edge reciprocates on the work piece and this action accomplishes the cutting of the work piece. Following are the different types of saws:

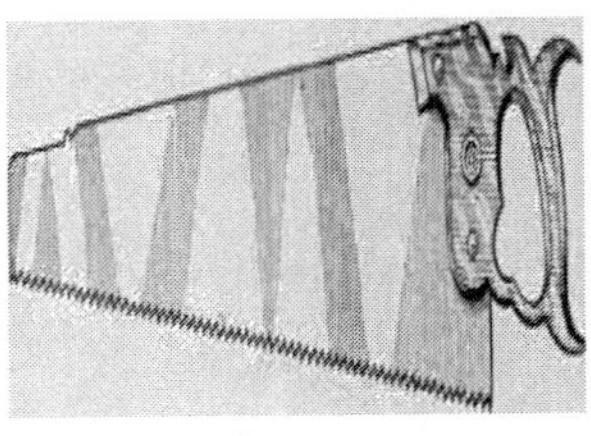

**Figure 16.** Rip Saw

- **Rip Saw:** It is used for cutting along the grain in a thick wood. The blade used in this saw is of high-grade steel. And its handle is made up of hard wood I which the blade is attached. The cutting edge makes an angle of 60 degrees with the surface.

The blades are of length nearly. The blades on this type of saw are almost 700 mm long and have 3 to 5 teeth per 25 mm.

- **Cross-Cut Saw:** It is also called a 'Hand Saw' and is used for cutting across the grain in thick wood. The length of the blade in this type of saw varies from 600-650 mm. in this saw, there are 8-10 teeth per 25 mm. The set of teeth in this type of saw is less as compared to the rip saw.
- **Tenon Saw:** It is also called 'Back Saw'. It is used for cutting small work. It is made up of a brass or steel back. The length of the blade varies from 250-400 mm and teeth per 25 mm are 13. The teeth are shaped in an equilateral triangle and are called peg teeth.
- **Panel Saw:** With the teeth per 25 mm varying from 10-12, it is nearly 500 mm long. It is very much like a cross-cut saw but has slightly more hook than a cross-cut saw.
- **Dovetail Saw:** It is a smaller version of a tenon saw. It has lengths varying from 200-350 mm, with the teeth per 25 mm, varying from 12-18. It is used for high accuracy.
- **Bow Saw:** It contains a narrow blade fitted in a steel or wooden frame. These saws are used for cutting curves as the handle can revolve in their sockets.
- **Coping Saw:** Just like a bow saw, it also has a blade of a similar kind and is held in a metal frame rigidly. It is used for cutting small curves.

**Figure 17.** Cross-Cut Saw

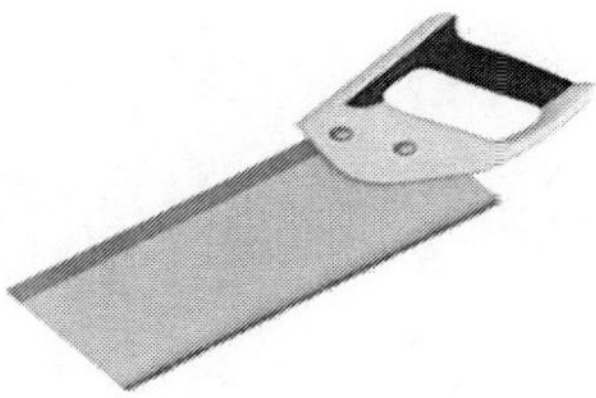

**Figure 18.** Tenon Saw

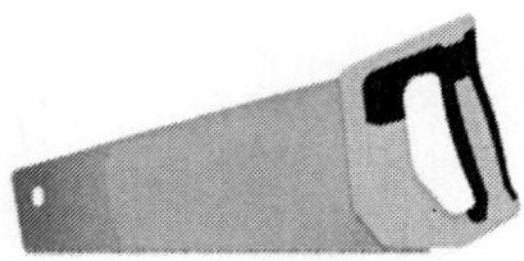

**Figure 19.** Panel Saw

**Figure 20.** Dovetail Saw

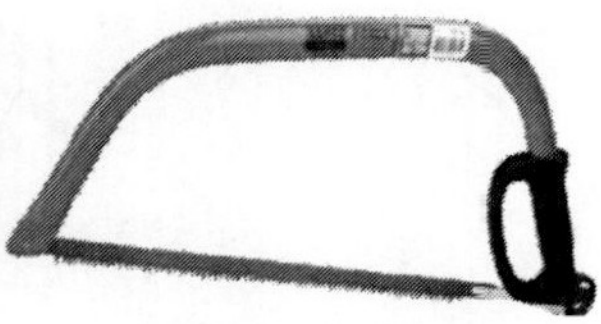

**Figure 21.** Bow Saw

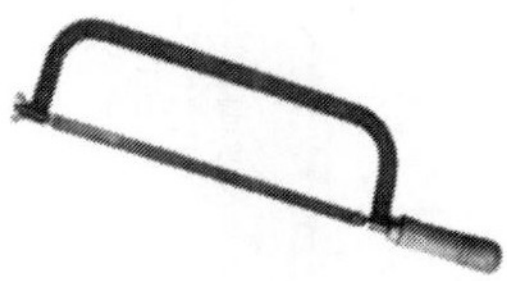

**Figure 22.** Coping Saw

- **Compass Saw:** It is used for heavy work, mostly used in radius cutting.

**Figure 23.** Compass Saw

i) **Chisels:** These are used for cutting wood accurately. There are a variety of widths for the blade, which varies from 3-50 mm. It is made of either forged steel or tool steel. It is designed as a wooden handle that is properly fitted with a steel shank.

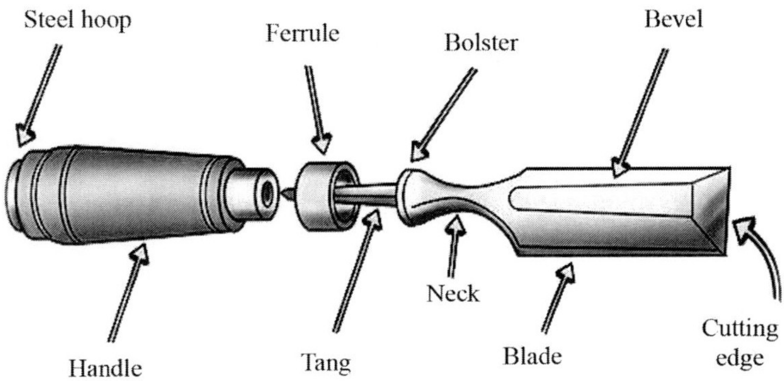


**Figure 24.** Chisel

## There are various types of Chisels

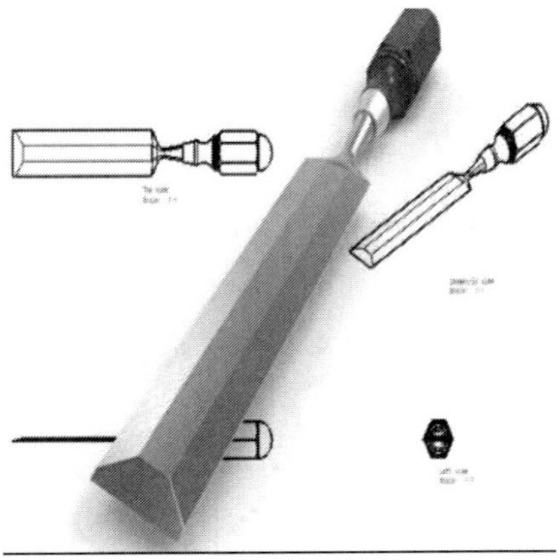

**Figure 25.** Beveled Edge Firmer Chisel

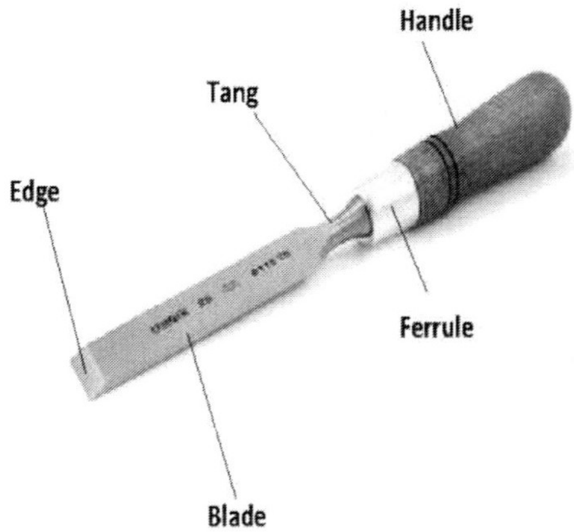


**Figure 26.** Firmer Chisel

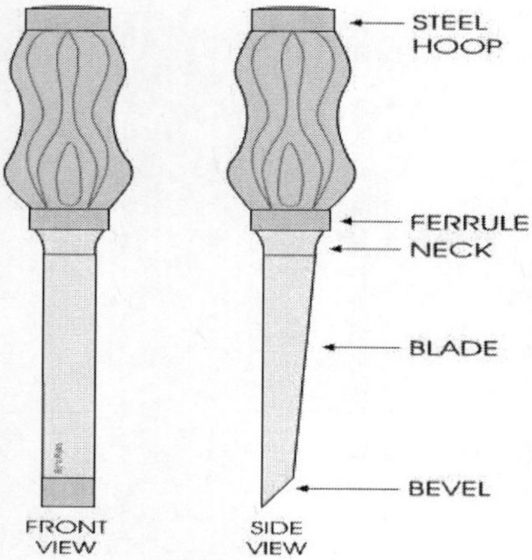


**Figure 27.** Dovetail Chisel

**Figure 28.** Mortise Chisel

## 4. Planning Tools

Planning tools are hand tools for producing flat surfaces on wood. In planning tools blades, we use are like chisels & we fit these blades in a block of wood or metal at some angle.

i) **Jack Plane**: It is the most common type of planning tool. It has a length of 35 cm. Its blade has a cutting edge of a slight curve.

ii) **Rebate Plane**: Used for marking rebates.

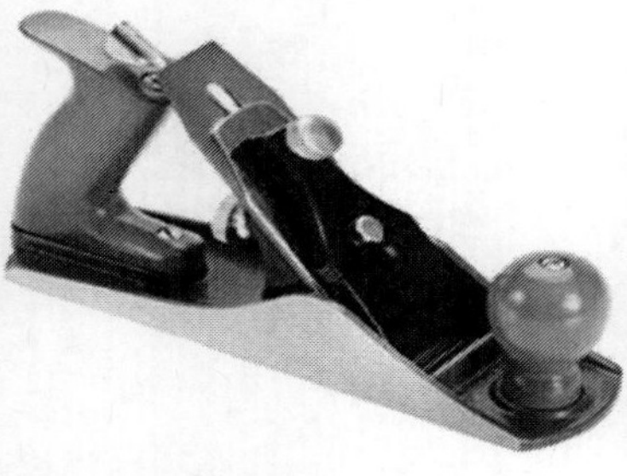

**Figure 29.** Jack Plane

**Figure 30.** Rebate Plane

iii) **Smoothing Plane:** It is used for surface finishing. The blade's edge is straight cutting & has a length of nearly 20-25 cm. It is better than the jack plane. When the jack plane is used, after that, we use this plane.

iv) **Plough Plane:** We use this plane for making grooves.

Figure 31. Smoothing Plane

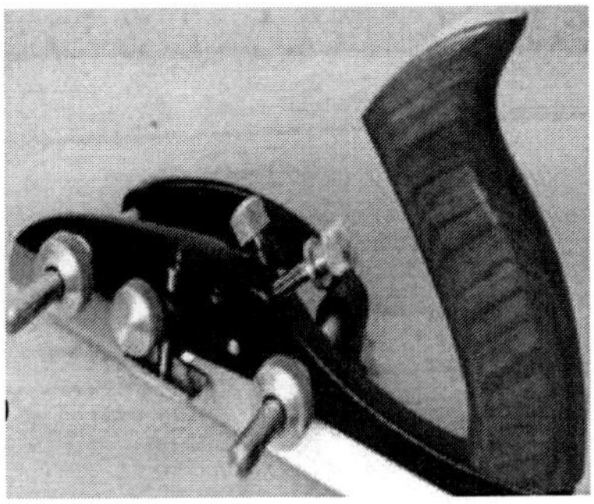

Figure 32. Plough Plane

## 5. Miscellaneous Tools

i) **Wood Rasp File:** It is used for finishing the surface. It has sharp cutting teeth.

ii) **Screw Driver:** These are used for driving the screw on the boards.

iii) **Pincer:** Its main work is to pull out nails.

Figure 33. Wood Rasp File

Figure 34. Screw Driver

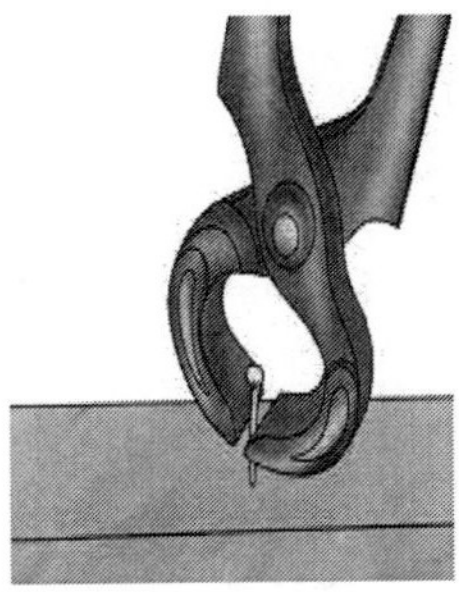

Figure 35. Pincer

## 6. Striking Tools

These are used for driving nails and operating a chisel.

i) **Warrington Hammer:** It is also called the 'Peen' hammer. It is mostly used for light work. It is made of cast steel.

ii) **Claw Hammer:** Driving of nails is done by the face and we can pull the nails out by claw, so it can be used for two purposes. It varies from 375 to 675 gms.

iii) **Mallet:** It is used for operating gauges and chisels. It is made up of hard wood with a handle. The faces that strike on the surface are made flat. The cross-section of the flat surface can be either square or triangular.

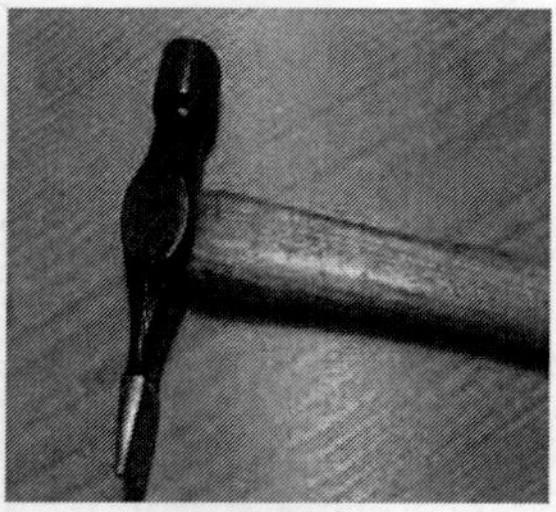

**Figure 36.** Warrington Hammer

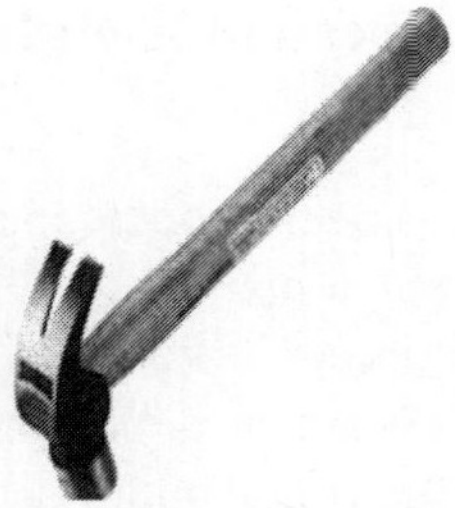

**Figure 37.** Claw Hammer

**Figure 38**. Mallet

### 7. Drilling and Boring Tools

These are used for making holes in wooden works.

i) **Gimlet:** It is a tool that is operated by hand and makes holes having small diameters for screws.

ii) **Hand Drills:** If we want to drill small holes, for that we can use this. It has a straight shank. It is small and light in weight.

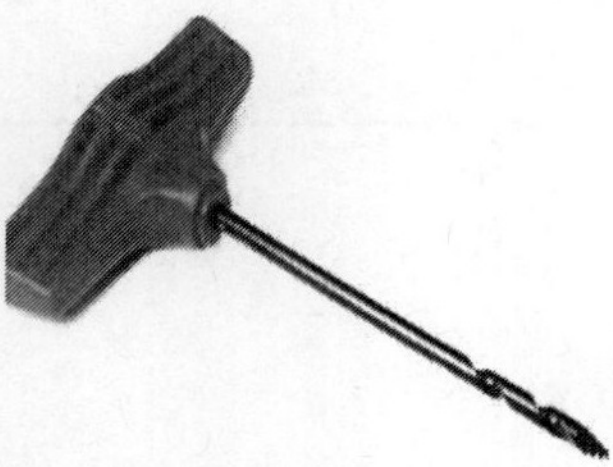

**Figure 39.** Gimlet

**Figure 40.** Hand Drills

iii) **Auger:** If we want to drill deep holes, we can use this. It is made by using a steel bar. Its top is provided with a handle and the bottom with a screw point.

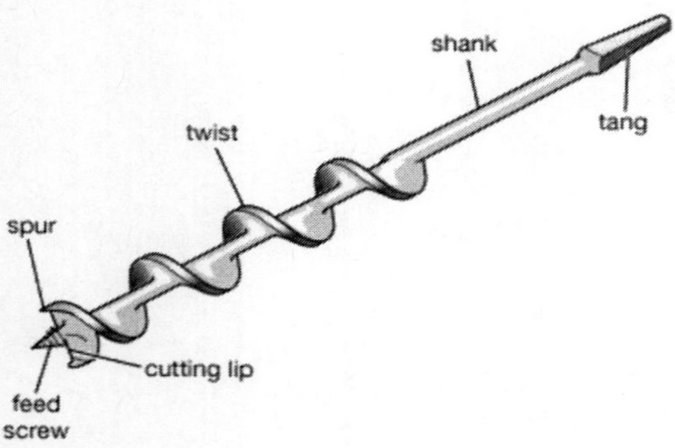


**Figure 41.** Auger

## Study of Various Types of Timber and Their Identification

### Introduction

Timber is a type of wood that has been processed and used as beams for house purposes, furniture making and boat making too. The other meaning means of the word Timber is Lumber and timber is wood or firewood of growing trees, or we can say it is hardwood. Timbers are used for structural purposes and designed in defined shapes for making furniture and building houses. Timber is the material used for carpentry. Timber is obtained from exogenous trees (having broad leaves). It is very useful for engineering and architecture purposes by sawing it into different sizes.

**Table 1:** Export and Import of Indian Timber in World

| Production Quantity (*1000 $M^3$) | Import Quantity (*1000 $M^3$) | Domestic Consumption | Exports Quantity | |
|---|---|---|---|---|
| Logs (INDRoundwood) | 49,514 | 4,383 | 53,881 | 19 |
| Sawnwood | 6,889 | 869 | 7,744 | 14 |
| Veneer | 295 | 415 | 702 8 | |
| Plywood | 2,537 | 141 | 2,627 | 51 |

The above-given information is from ITTO (2019).

### Now Export Market of India in the World (Pie chart):

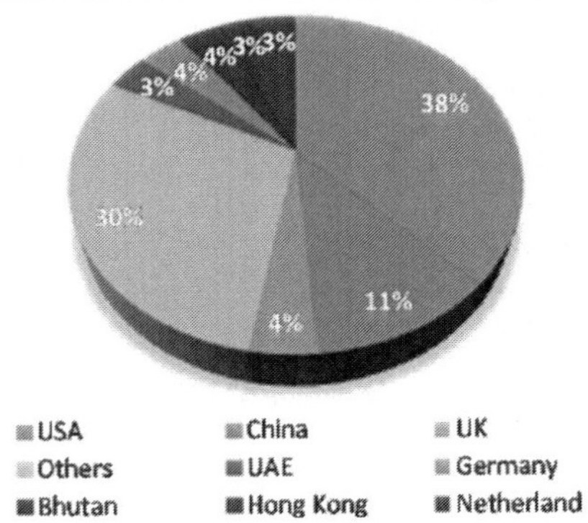


**Figure 42.** Pie Chart of Export Market of India

### Properties of Timber

1. It is easily available.
2. Lighter and stronger to use.
3. It is good for polishing and painting, according to us.
4. Suitable for soundproof constructions.
5. It is very economic.

**Now we see some details regarding Timbers and their Identifications:**

**1. Shisham**

**Figure 43.** Shisham Tree

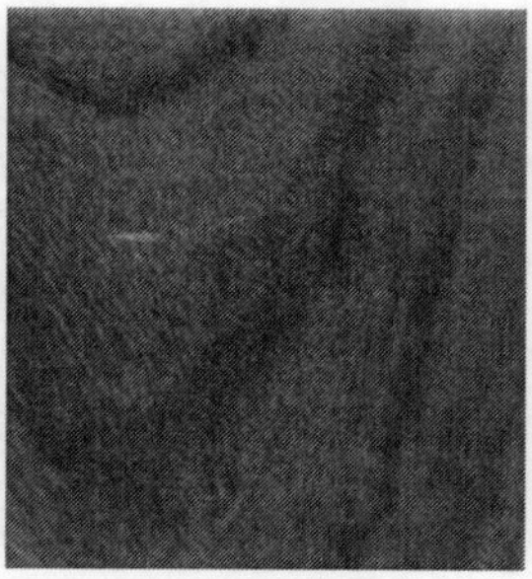

**Figure 44.** Cross Cut

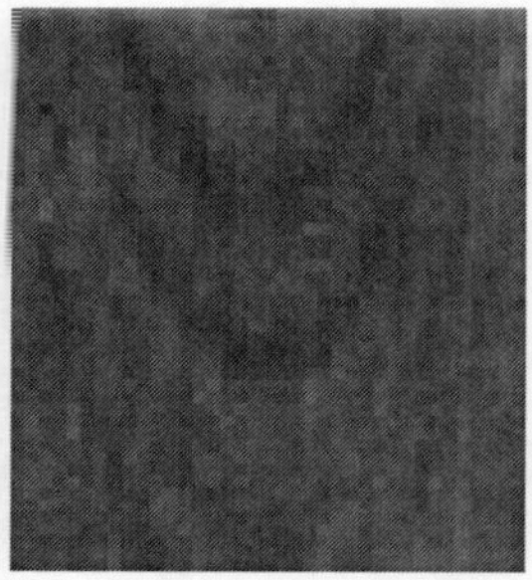

**Figure 45.** Flat Swan

**General Features:** The heartwood of Shisham is found golden brown to dark brown with highly dark strips in it.

**The following are to be checked for identification:**

Hard and moderately heavy wood. Vessels are massive and, mostly filled with dark and gummy deposits.

| | |
|---|---|
| Colour | • Golden brown to dark brown.<br>• Deep dark streaks. |
| Weight | Moderately heavy. |
| Grain | Narrowly interlocked. |
| Texture | Medium to coarse. |
| Strength | Strong. |
| Durability | • Very durable.<br>• Highly termite resistant. |
| Treatability | Extremely resistant. |
| Working Properties | Easy<br>• Planning.<br>• Boring.<br>• Turning.<br>Good<br>• Nailing (necessary: pre-boring)<br>• Finish. |
| Uses | The best timber for cabinetry and furniture<br>• Boat<br>• Agricultural and musical instruments<br>• Marine and air craft grace plywood<br>• Decorative plywood<br>• Block boards<br>• For tobacco pipes, root wood is used. |

## 2. Teak

**Figure 46.** Teak Tree

**Figure 47.** Cross Cut

**Figure 48.** Flat Swan

## General Features

- It is a heart wood with golden brown in color with black strips.
- Wide color variation.
- Oily feel.
- Smell like old leather.
- Ring porous hard wood.

### Following are to be checked for identification:

- Wood is moderately hard and moderately heavy.
- Vessels are filed.

| | |
|---|---|
| Colour | • Heartwood is golden brown when fresh<br>• Dark brown on exposure<br>• Waxy feel<br>• Sapwood is white, pale yellow or grey |
| Weight | • Hardl heavy (650 kg/m$^3$ at 12% m. c) |
| Grain | • Mostly Straight<br>• Sometime Wavy |
| Texture | Coarse |
| Strength | Strong |
| Durability | Durability: high |
| Treatability | Resistant: extremely |
| Working Properties | Easy:<br>• Planning<br>• Boring<br>• Turning. With pre-boring, nailing: good<br>• Good: finish |
| Uses | • It is a very versatile wood<br>• Extensively used for ship and boat building<br>• Due to the acid and alkalis-resistant properties, it is a very good choice for laboratory accessories<br>• Stand for ammunition explosive boxes |

## 3. Mango

**Figure 49.** Mango Tree

**Figure 50.** Cross Cut

**Figure 51.** Flat Swan

### General Features

- Heavy and hard wood
- Wood: yellowish-white to grayish-brown wood

### Following are to be checked for identification:

- Growth rings are fairly distinct.
- Wood is moderately hard and heavy.
- Fine to moderately broad rays.
- Vessels are in decreasing order (large to medium), and also filled with tyloses, either large in number in a curve like multiples 2-3.

| | |
|---|---|
| Colour | Yellowish white to greyish brown. |
| Weight | Moderately hard and moderately heavy.690 kg/m3 at 12% m.c. |
| Grain | Straight to somewhat interlocked. |
| Texture | Medium to coarse. |
| Strength | Strong. |
| Durability | Non-durable. |
| Treatability | Easily treatable |
| Working Properties | Also easily to saw, satisfies the machining work and polishing over it is best. The holding capacity of the nail and screw is much better. Also satisfies the wood peeling property. |
| Uses | • Plywood<br>• Cabinets<br>• Boat and ship. |

## 4. Himalayan Fir

**Figure 52.** Himalayan Tree

**Figure 53.** Cross Cut

**Figure 54.** Flat Swan

**General Features:** Is soft and a plank of light wood.

**Following are to be checked for identification:**

- These woods are turned into a light brown color from the creamy white color as on their aging, these have no heartwood or special odor.
- The growth rings of these woods are usually gradual from their transition state (early to latewood).
- The rays are very fine, only recognizable as fine lines lying close together under a magnifying glass.
- Resin canals are absent.

| | |
|---|---|
| Colour | • It is creamy white<br>• On aging, it turns to light brown<br>• No odor. |
| Weight | Nearly 450 kg/m$^3$ at 12% m.c. |
| Grain | Straight |
| Texture | Fine and even |
| Strength | Weak |
| Durability | • It gets decayed very easily in contact with the ground<br>• It is durable indoors. |
| Treatability | It is easy to see. |
| Working Properties | Good finish. |
| Uses | In high demand, it is often used for the purpose of light packing (tea bag or fruits case). |

## 5. Haldu

**Figure 55.** Haldu Tree

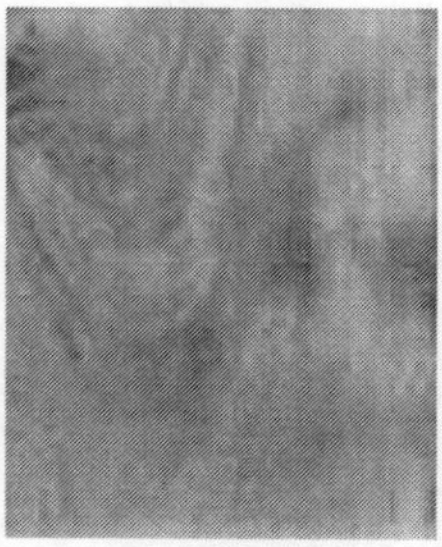

**Figure 56.** Cross Cut

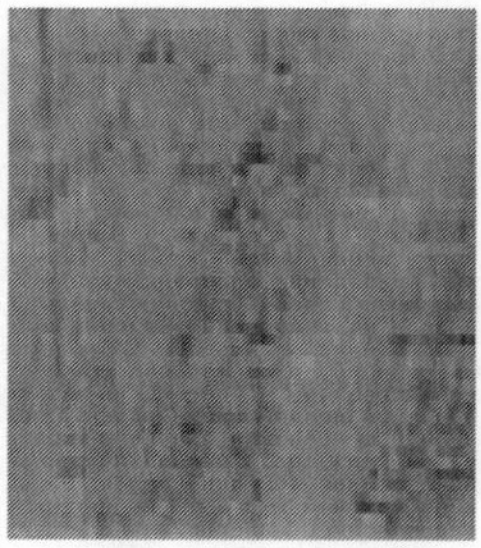

**Figure 57.** Flat Swan

**General Features:** Deep yellow color.

**Following are to be checked for identification**

- When these woods are exposed, then their color turns from yellow to reddish/ brownish.
- Generally, these woods are light to moderate-heavy and with a fine texture.
- Parenchyma is not seen clearly by the naked eye.
- Rays are fine, numerous and closely spaced.
- Vessels are small, single, or in the multiple of two to three curves.

| | |
|---|---|
| Colour | It is found that the sapwood is paleyellowish or yellowish-white. |
| Weight | Light-weighted as well as heavy, bothkinds of wood are found. (695 kg/m$^3$ at 12% m.c). |
| Grain | Generally, interlocked. |
| Texture | Generally, fine. |
| Strength | It's strong. |
| Durability | It is not that durable. |
| Working Properties | • Sawing can be performed easily<br>• With satisfactory machining<br>• Provides a good finish. |
| Uses | • Carving wood is the best type for this purpose<br>• Plywood of class 1 is made by the use of this<br>• Battery separators. |

## 6. Babul

**Figure 58.** Babul Tree

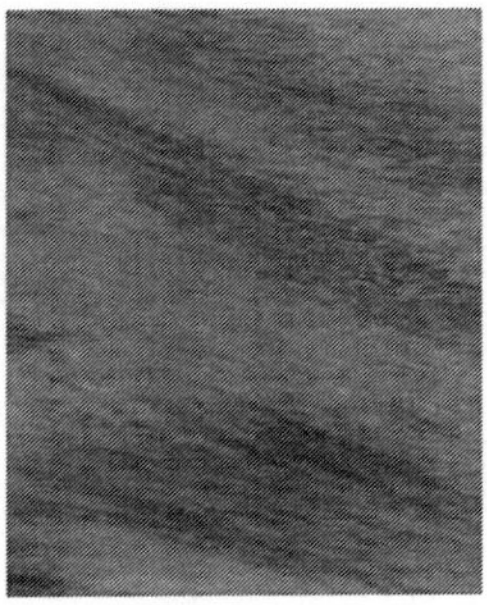

**Figure 59.** Cross Cut

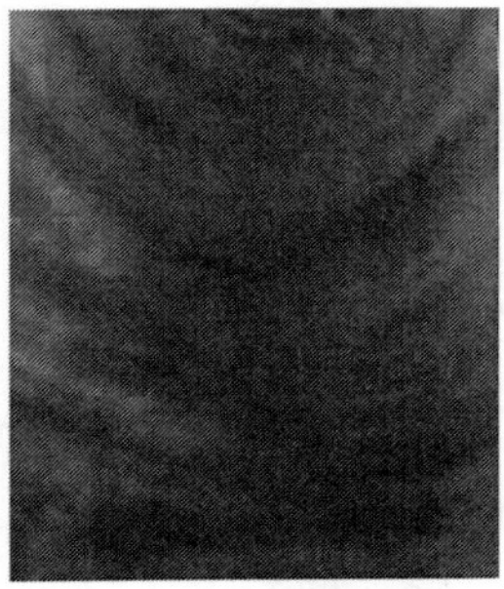

**Figure 60.** Flat Swan

**General Features:** The heart wood varies from pinkish-brown to reddish-brown.

### Following are to be checked for identification:

- Wood is hard and heavy.
- Growth rings are indistinct.
- It has a few small vessels and dark brown gummy is filled in it.

| | |
|---|---|
| Colour | Sapwood varies from white to pale yellow. |
| Weight | Hard and heavy (800 kg/m$^3$ at 12% m.c). |
| Grain | Straight to interlocked. |
| Texture | Medium to coarse. |
| Strength | Strong. |
| Durability | Very durable. |
| Treatability | Resistant. |
| Working Properties | Easy<br>• Planing<br>• Boring<br>• Turning.<br>• Good: Nailing but requires pre-boring Finishing is also good. |
| Uses | • For constructing posts, rafters, and bridges<br>Most importantly, this tree produces a pale yellow to brown gum that is widely used in confectionery and medicine. |

## 7. Banyan

**Figure 61.** Banyan Tree

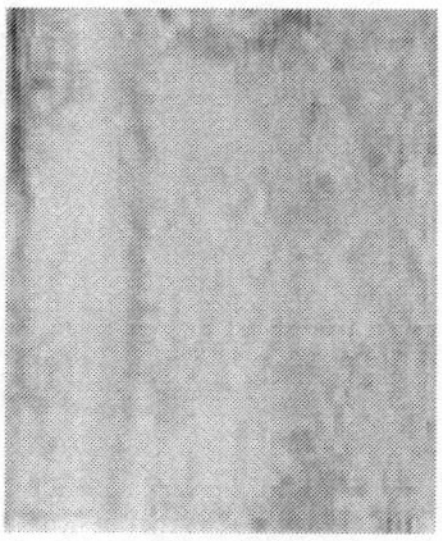

**Figure 62.** Cross Cut

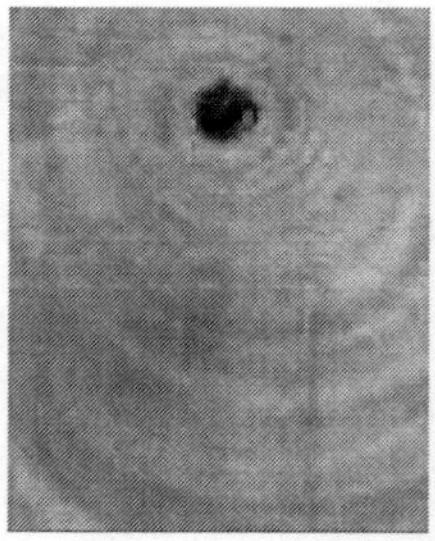

**Figure 63.** Flat Swan

**General Features:** Varies from light to moderately heavy wood.

| | |
|---|---|
| Colour | It turns from creamy white to greyishwhite when exposed and with its aging, it turns to grey and pale brownish in color. |
| Weight | Light to moderately heavy. Air-dryspecific gravity approx. 0.61 |
| Grain | Interlocked. |
| Texture | Uneven. |
| Strength | It is not strong enough. |
| Durability | Perishable. Timber does not exist in an exposure atmosphere (moisture) but it exists in either dry places or in a lessdurable atmosphere. |
| Treatability | Easy. |
| Working Properties | Sawing: it is very easy to get a fleeting-like figure (with the help of flat sawing); a mottling figure may be obtained by flatsawing. |
| Uses | Low quality (3rd class) wood is used for making toys, cartoon boxes or very light packings. The wood of aerial roots (roots above theground) is used for carrying shafts, supporters of tents, etc. |

## Practice of Marking, Planning, Sawing, and Chipping in the Carpentry Shop

### Tools required

- Chisel
- Wooden mallet
- C clamp
- Try square
- Rule
- Jack planner.
- Handsaw.
- Marking gauge.
- Bench vice.
- Pencil.

## 1. Marking

**Figure 64.** Marking

Marking is done on the workpiece to get an accurately finished product. In this set of dimensions, a workpiece can be done. The procedure of the marking process is as follows :

i) Firstly hold the work in the holding device depending on different shapes and sizes.

ii) For flat workpieces, use the surface plate and for round workpieces, use the clamp.

iii) With the help of the ruler and the square, a fine, narrow line can now be drawn with a pencil to mark the desired dimensions. Marking is also done with the help of a gauge.

iv) For accuracy, we can also use blades for marking.

v) Divider & compass is used for marking arc and circle.

## 2. Planning

**Figure 65.** Planning

Planning operation is done to produce a flat surface or to smoothen the wooden surface. The procedure of planning :

i) Firstly place the workpiece in the vice.

ii) Then marking can be done with the help of a marking gauge.

iii) Then starting from the first edge, plane the total length of the workpiece with the help of a metal jack plane.

iv) Then check the trueness of work by try- square and looking below to find the high and low spots.

## 3. Sawing

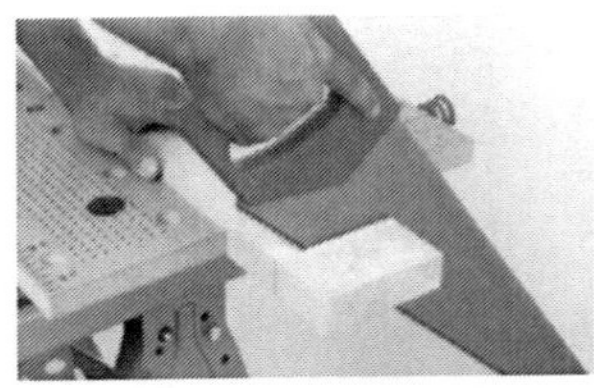

**Figure 66.** Sawing

In the sawing operation, the workpiece is cut into different shapes and sizes with the help of a hacksaw. The procedure for sawing operation is as follows:

i) First, the workpiece is fitted into the carpentry vice.

ii) Now we make slots with the use of a try square and pencil.

iii) Place the right hand on the handle of the saw and the left hand near the marking line of the workpiece, position the saw at a 45° angle and keep straight.

iv) And then, move the saw back and forth along the marking line to get the required cut.

## 4. Chipping

In chipping, operation removal of material is performed by using a tool named a chisel. Extra materials are removed by striking the chisel on the wooden workpiece with a hammer. The procedure of chipping is as follows:

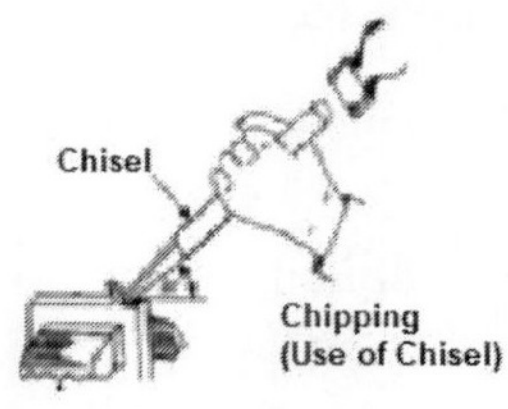


**Figure 67.** Chipping

i) First, marking is done on the wooden workpiece to be cut.

ii) Held the workpiece in a vice.

iii) With the left hand, hold the chisel and with the right hand, hold the hammer.

iv) Then place the chisel on the marked line and strike the hammer on the head of the chisel.

v) And during the process, we should concentrate on its cutting edge.

**Precautions**

1. Always wear goggles and a safety mask during sawing and chipping operations.
2. Wood pieces should be free from moisture.
3. Always start slowly to prevent buckling of the handsaw.
4. Apply grease on the edge of the chisel during the chipping process.
5. Mushroom head chisel or long chisel never be used for chipping.

## Experiment 1

**Aim:** Preparation of simple joints: Making of Cross Half Lap Joint.

### Tools Required

- Bench vice.
- Bevel-edged chisel.
- Tenon saw.
- Scriber.
- Marking knife.
- Rip saw.
- Wood rasp file.
- Jack plane.
- Steel rule.
- Try square.
- Marking gauge.
- Pencil.
- Mallet.
- Hand saw.

**Material Required:** Piece of teak wood of size 50 × 150 × 300 mm.

### Theory

- This type of joint is easy to make & it is strong as well.
- We use this joint for creating continuous and beautiful lines in the furniture.

### Sequence of Working

- Planning.
- Marking and measuring.
- Sawing.
- Chiseling.
- Finishing.
- Joining.

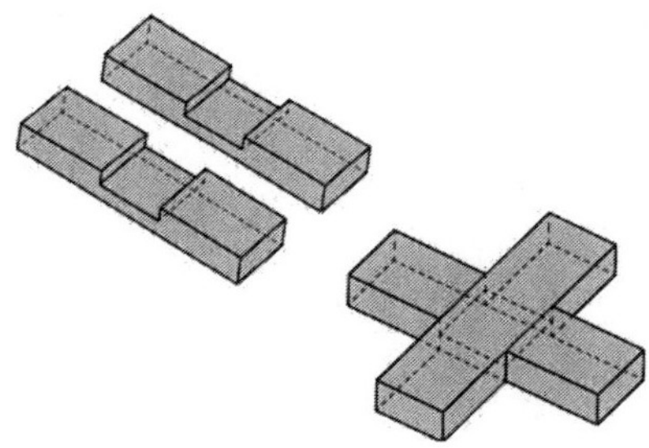

**Figure 68.** Cross Half Joint

### Procedure

1. Cut the wooden piece into two halves of identical size using a hand saw and check for dimensions.
2. Plane one side of the work with a jack plane and check for straightness.
3. Plane another side and check for squareness with a try square.
4. Mark required dimension lines with a marking gauge.
5. Plane the extra material to the correct size.
6. Now fix the work piece on the vice & saw along the lines marking the width of the socket.

7. Make a series of cuts into the waste wood.
8. The excess material is chiseled for both parts with a mortise chisel.
9. Finish the chiseled part.
10. Join the pieces using a mallet to get the joint.
11. Joint them by using glue.
12. Now let the joint dry.

**Result:** Cross-Half joint is made successfully as per the desired dimensions.

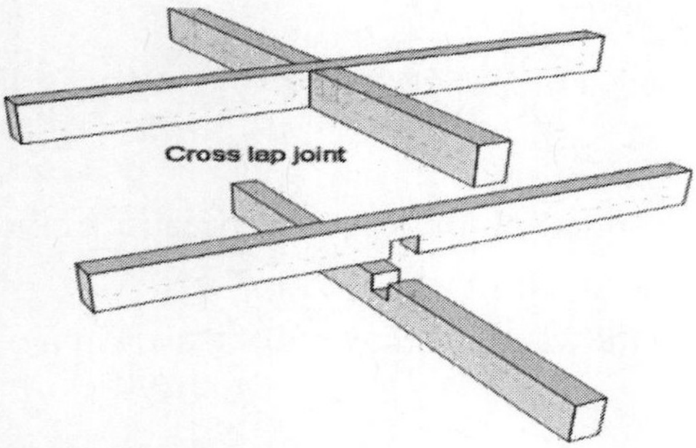


**Figure 69.** Cross Lap Joint

## Precautions

1. The workpiece should be free from moisture.
2. Loose clothes are to be avoided.
3. Always wear goggles and a safety mask during sawing.
4. During planning, uniform pressure should be applied.
5. Use the most effective sharp tools.
6. Hands ought to now no longer be located in the front of the sharp-edged gear.
7. There should be no parallax error in marking.
8. Chisel carefully.
9. The joining of pieces should be tight.

## Experiment 2

**Aim:** Preparation of simple joints: Making of T-Halving joint.

### Tools Required

- Carpenter's vice
- Jack plane
- Tenon saw
- Firmer chisel
- Marking gauge
- Wood rasp file
- Pencil
- Try square
- Steel rule
- Hand saw
- Scriber
- Cross-cut saw
- Mallet
- Marking knife

**Material Required:** Pieces of soft wooden of size 50 × 50 × 300 mm.

### Theory

- This type of joint is weak and it can split. Splitting happens because of the absence of shoulders.
- This joint is mainly used for parting the trays and drawers of a table or any furniture.
- It is also used for making torsion boxes.

### Sequence of Working

- Measuring and marking
- Planning
- Squareness check
- Joining
- Excess material removal
- Sawing
- Chiseling
- Finishing

### Procedure

1. Check the work piece for size.
2. After checking the work piece, it is clamped in the carpenter's vice. Then any two surfaces are planned by using a jack plane and checked by try square for squareness.
3. The marking gauge is set and mark 44 mm on the width, then highlight the mark using a pencil.
4. Chisel unwanted material.
5. After chiseling, it is planned.
6. Mark 22 mm on the thickness side using the marking gauge and steel rule.
7. Mark 120 mm on the piece using the steel rule and try square.
8. After that, cut the work into two parts measuring 120mm using a cross-cut saw.

9. Using the steel rule and try to mark the square at a distance of 44 mm, 38 mm, and 44 mm on one piece.
10. Cut the piece with a hand saw and chisel it.
11. Mark a 22 mm distance on the thickness side at 44 mm from one end of the 120 mm workpiece, then cut it by using a handsaw and chisel it.
12. Mark a distance of 44 mm on one end of the second piece and 11 mm on the thickness side by using a steel rule and try square, then cut by using a hand saw and chisel it.
13. Remove all the unwanted portions of wood with a firmer chisel.
14. The final finishing is done with a rasp file.
15. Join the pieces through glue using a mallet to obtain a slightly tight joint.
16. Now let the joint dry.

**Result:** T-Halving joint is formed as per the specified dimensions.

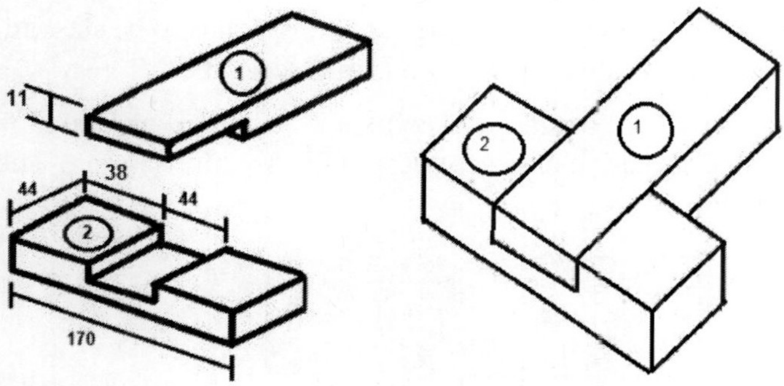


**Figure 70.** T-Halve Joint

## Precautions

1. Marking should be done carefully.
2. Wood pieces should be free from moisture.
3. Avoid wearing loose clothes.
4. Place tools in their correct positions.
5. Don't put your hands in front of a sharp tool.
6. Uniform pressure should be applied during planning.
7. Cut the work piece carefully with a chisel.
8. Handle sawing and planning tools carefully.
9. Wooden items with nails ought never to be allowed to stay on the ground.
10. Keep the screwdriver properly pointed to stop injuring to hand.

## Experiment 3

**Aim:** To prepare a Dovetail Lap Joint.

### Tools Required

- Carpenter's wise.
- Jack plane.
- Steel rule.
- Try square.
- Marking gauge.
- Firmer chisel.
- Cross out saw.
- Scriber and mallet.
- Tenon saw.

**Material Required:** 50 mm * 50 mm * 250 mm wood pieces.

### Procedure

### 1. Lay-outing

i) **Marking: S**pecimen marks perfectly for that pencil are used. After this, we use a knife is used for the sectioning, the section which is to be slashed, and then fine lines are generated on the wood fiber sections.

ii) **Scribe the Lap:** The width of the lapped segment is remarked by using the tool gauge. Its scribed traces will assist and guide us while sawing. The gauge set should be afterward.

**Figure 71.** Marking

**Figure 72.** Scribe the Lap

### 2. Begin with a hand-cut half lap

**Figure 73.** Hand-Cut Half Lap

i) Firstly backsaw of a fine blade is used to cut/ slash the beginning half lap. Taking some initial strokes and the waste side has carefully stayed. In the case of imperfect cuts, when the sawing is done, we can remove excess material with the use of a wide chisel.

ii) **Making a 3-Stage Cut:** The workpiece is angled away from us, and the saw horizontally is maintained as we slash and stop at the short of the shoulder's mark. Then working slowly and precisely, we will pivot the piece toward us and repeat the maneuver. At last, the piece is set to a vertical position, and the saw is played.

**Figure 74.** Making a 3-Stage cut

iii) **A Guiding Notch:** We start the shoulder cut for your saw by making a guide notch.

A fine chisel is held at some angle and gently we remove material nearly 1D 32 inches, following the lay line as reference.

iv) **Sawing the Shoulder:** Once again, the saw shoulder down to cheek for staying on the waste side of the line.

**Figure 75.** Sawing the Shoulder

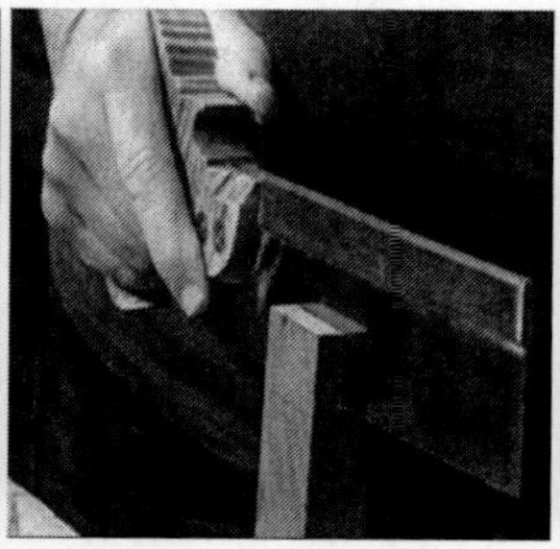

**Figure 76.** A Guiding Notch

**Figure 77.** 3-Stage Cut

### 3. Make the tail

i) We will do marking on the basis of what kind of face we want. For marking, we can use shop-made dovetail gauges, bevel gauges or straight edges.

ii) **Laying Out the Tail:** Basically, dovetails are marked out with the help of a tool, a simple gauge. Then after marking out the face, a square of small size is used to increase the length of the line from the end of the specimen to guide our saw cuts.

**Figure 78.** Laying Out the Tail

iii) **Saw the Shape**: The specimen is holding in the vise within the cutline, which is normal to the bench through which the precision is increased and then we saw the tail edge in turn. Special care is taken while sawing squarely from front to back. After that, the square of small size is used to detect that every edge is squared to the face of the tail, which is a very important step for the well-fitted joint.

Figure 79. Saw the Shape

## 4. Layout the socket

i) **Gauging the Length of the Tail:** The length of the tail is marked with the help of a gauge. Now the second marking gauge is set to confirm the correct length of the dovetail.

ii) **Transfer the length:** We use a set gauge to scribe the length of the socket which is measured from the tail, on the wood piece for making the socket.

Figure 80. Gauging the Length of the Tail

iii) **Trace the Tail Slope:** With a knife or a very sharp pencil, we trace the tail angles. The lines extended at the neck of the tail squarely across the adjacent faces then the depth of the socket is marked on that face by using the tool gauge.

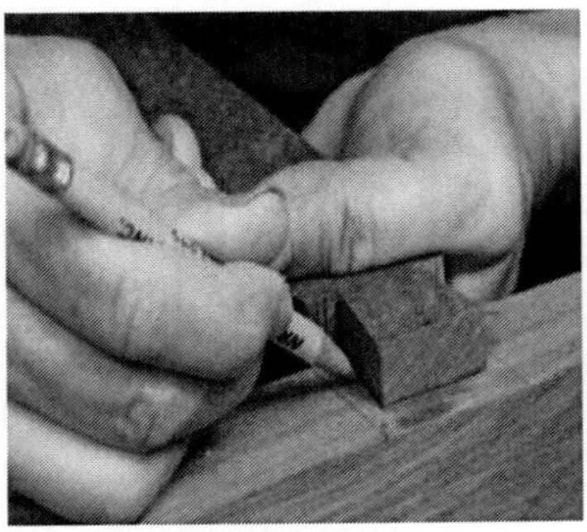

Figure 81. Trace the Tail Slope

Figure 82. Transfer the Length

## 5. Saw and chop the socket

i) **Sawing the Sides:** While holding the saw at 45°, we cut the portion inside the tail's lay out lines.

Figure 83. Sawing the Sides

ii) **Chopping:** To chop out or remove the waste in layers from the socket's bottom. As we tap with our mallet, we should angle our chisel blade slightly upward for easy cutting. Then we stop it just before the line of depth and then restart the work on the socket side. While carefully holding the chisel vertically, tap downward until we have removed all of the waste in the socket to just a bit inside our layout lines.

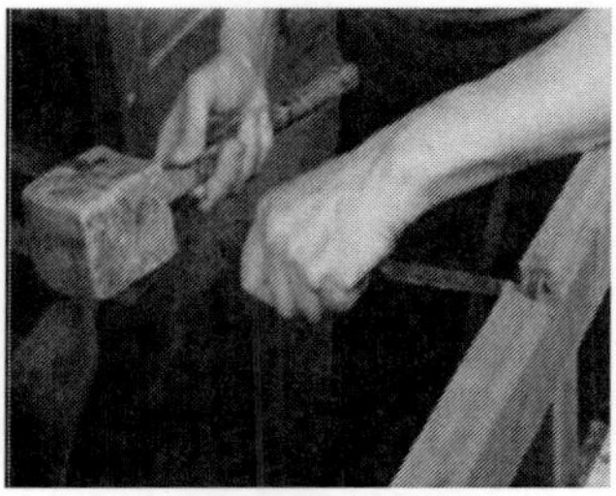

**Figure 84.** Chopping

## 6. Pare to fit

i) **Leaning-In:** For a good fit, it's okay to undercut the socket sides a little.

ii) **Paring:** Paring is the process of removing the rough surface from the specimen by the use of a saw or chisel. This operation is performed to make the wood surface regular.

iii) **Fitting:** There should be an ideal fit, just require a little light hammering.

**Figure 85.** Leaning-in

**Figure 86.** Paring

**Figure 87.** Fitting

**Result:** The dovetail joint is formed as per the specified dimensions.

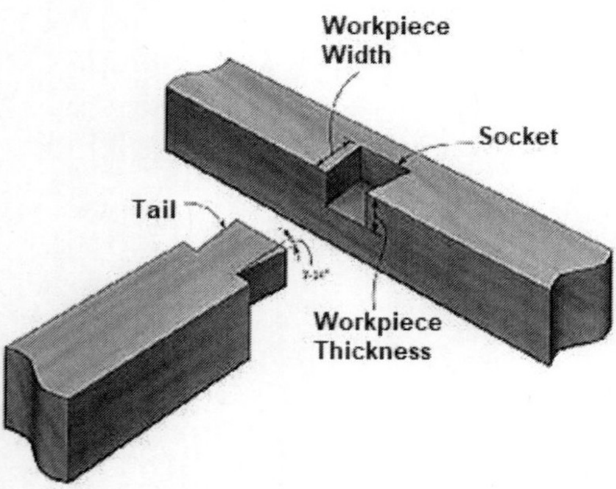


**Figure 88.** Dovetail Joint

## Precautions

1. The wooden pieces must be fitted between the carpentry properly.
2. Uniform pressure should be applied while planning.
3. Chisel must be done layer by layer so that we get a finished product.
4. Be careful not to remove too much material from the socket's sides while leaning in.

## Experiment 4

**Aim:** Preparation of simple joints: Making of Mortise and Tenon Joint.

### Tools Required

- Carpenter's vice.
- Jack plane.
- Tenon saw.
- Marking knife.
- Scriber.
- Ratchet brace.
- Mallet.
- Hand saw.
- Depth gauge.
- Steel rule.
- Mortise chisel.
- Try square.
- C-clamp.
- Marking gauge.
- Pencil.
- Adze.

**Material Required:** Pieces of teak wood of size 35 × 45 × 300 mm.

### Theory

- This type of joint is very strong and durable.
- But another fact is that the variation of temperature and humidity results in little expansion or contraction of wooden members.
- It calls for massive capacity to make. It is typically applied within side the improvement of cost furniture.

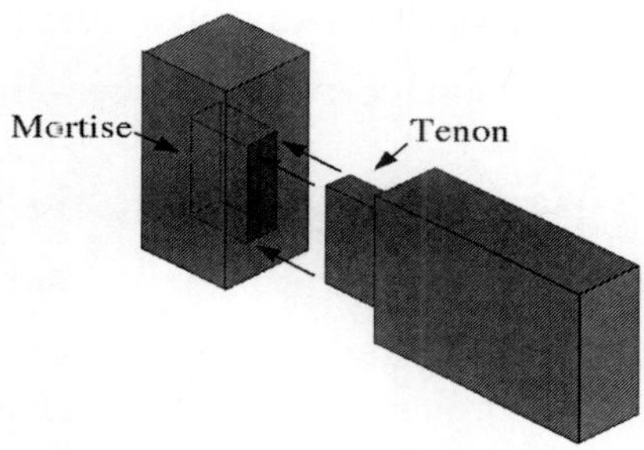


**Figure 89.** Shape of Mortise and Tenon

### Sequence of Working

- Marking and measuring.
- Planning.
- Squareness checking.
- Excess material removal.
- Finishing.
- Sawing.
- Setting.
- Chiseling.
- Joining.

### Procedure

#### 1. Laying out the joint

i) Take two pieces of wood, and mark the depth on the tenon piece that you want to provide to the joint by placing the mortise piece over the tenon piece.

ii) Fix the marking gauge on the vice and set it to the required depth.

iii) Fix the tenon piece in the vice and with the marking gauge, mark one shoulder line on one side.

iv) Continue the shoulder mark on the remaining sides & highlight them using a pencil.

v) Attach the tenon piece onto the vice and mark the cheeks of the tenon on the end grain and mark the width of the tenon on the end grain.

**2. Cutting the Tenon**

i) Place the tenon piece in the vise horizontally and cut the shoulder line across the wide face to where it meets the cheeks.

ii) Turn the piece over and repeat the step on the opposite side.

iii) Now fix the tenon piece vertically on the vice and sawed down the remaining two sides until the cheeks.

iv) Leave some extra wood for the finish.

v) Smooth the sawed side by using a chisel.

**3. Marking & Cutting the Mortise**

i) Mark the mortise piece according to the previously made tenon.

ii) For shaping, the mortise bore out a series of holes with a ratchet brace according to the length of the tenon.

iii) Ensure smooth and square sides.

**4. Fastening the Joint**

i) Now fit the tenon into the mortise using a mallet & use a try square to check the squareness.

ii) Glue the joint and using a scraper, remove excess glue from the wood surface to be joined.

iii) Now let the joint dry.

**Result: Mortise and Tenon** joint is prepared successfully as per the required dimensions.

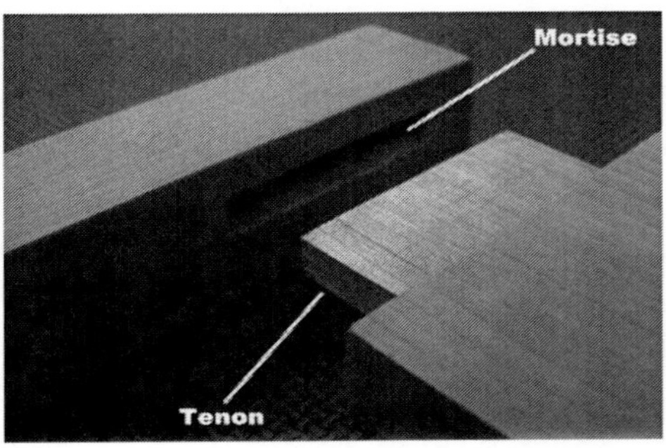

**Figure 90.** Mortise and Tenon Joint

## Precautions

1. Marking should be done carefully.
2. Wooden pieces should be free from moistur e.
3. Avoid wearing loose clothes.
4. Use only sharp tools during the carpentry process.
5. There should be no parallax error in marking.
6. Uniform pressure should be applied while planning.
7. Chisel must be done layer by layer so that we get a finished product.
8. The joining of pieces should be tight.
9. Handle sawing and planning tools carefully.

## Carpentry Shop Objective Questions

1. Which of the following is not a marking tool?

   a) **Chisel** b) Bevel
   c) Square d) Scribing Knife

2. Bevel is used for

   a) Setting Right Angles
   b) **Setting Angles other than a Right angle**
   c) Cutting Wood
   d) Cutting Timber

3. Which of the following is not a cutting tool?

   a) Compass Saw b) Firmer Chisel
   **c) Plier** d) Coping Saw

4. Which of the following is not a boring tool?

   a) Ratchet Brace **b) Screw Driver**
   c) Auger Bit d) Pointed Awl

5. The length of a spike is ———— .

   a) 5-10 cm **b) 10-15 cm**
   c) 15-20 cm d) 20-25 cm

6. Nails are in the ———— section.

   a) Rectangular b) Square
   **c) Trapezoidal** d) Circular

7. Which of the following is not a planning tool?

   a) **Rose Plane** b) Bead Plane
   c) Jack Plane d) Rebate Plane

8. Dovetail saw is used to cut the ———— .

   a) **Timber** b) Wood
   c) Hardwood d) Plywood

9. Which of the following cutting tools is used for cutting wood?

   a) Coping Saw b) Cross-Cut Saw
   **c) Compass Saw** d) Tenon Saw

10. Which of the following boring tool is used for deep boring?

    a) Centre Bit **b) Auger**

    c) Gimlet d) Brad Awl

## Carpentry Shop Subjective Questions

1. Differentiate between hard wood and soft wood.
2. What are the different joints used in wood working?
3. What is the specific use of wood-turning lathe?
4. Sketch and describe the following joints.

    a) Mortise and tenon joint b) Lap dovetail joint

    c) Bridle joint d) Rafter joint

    e) Notching joint

# 2

# Fitting Shop

## Introduction

Machine tools are such necessary tools or capable tools for producing work at a faster rate. But sometimes we need other things such as a bench for some occasion, or sometimes the need for other components is much more so that we can replace or repair things more accurately. For this, we use some amount of hand fitting. The machine tools like gauges etc., need some kind of bench work. The precise work depends upon the efficiency of the fitting tools or machines.

**Bench Work:** deals with the work product of components with the help of hands.

**Fitting Work:** refers to assembling parts and removing metal to get or make the desired grip.

"Fitting Shop " is the joining of different parts with each other in such a way that they will fix with each other."

Fitting operations make use of a lot of tools that can either be fitted into a machine or can also be operated by hand.

## Tools Used in Fitting Shop

- Work Holding Devices /Clamping tools.
- Cutting tools.
- Drilling tools.
- Threading tools.
- Measuring and marking tools.
- Striking tool.

**Figure 91.** Fitting Table

## 1. Work Holding Devices / Clamping Tools

i) **Work Bench:** Important operations in the Fitting shop are carried out on a table (called Work Bench). It is a holding tool to hold several components on its surface. It provides much support to the performer while performing several operations.

ii) **Bench-Vice:** It has a cast-iron body and cast-iron Figure51}. Workbench jaws. They are wear-resistant. The two jaws hold the workpiece in between them. Knurling is done to increase the gripping power.

iii) **Pipe Vice:** It is similar to bench-vice, but it is designed in some special manner, so cylindrical specimens are only fitted between the jaws of the pipe vice. It is also known as a plumbing tool.

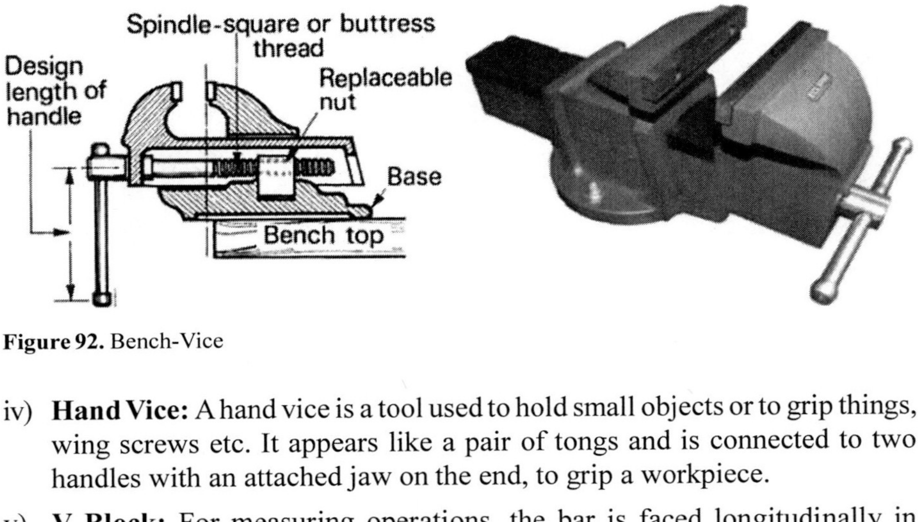


**Figure 92.** Bench-Vice

iv) **Hand Vice:** A hand vice is a tool used to hold small objects or to grip things, wing screws etc. It appears like a pair of tongs and is connected to two handles with an attached jaw on the end, to grip a workpiece.

v) **V Block:** For measuring operations, the bar is faced longitudinally in V Groove. Then the screw-on V clamp was clamped tightly.

**Figure 93.** Pipe Vice

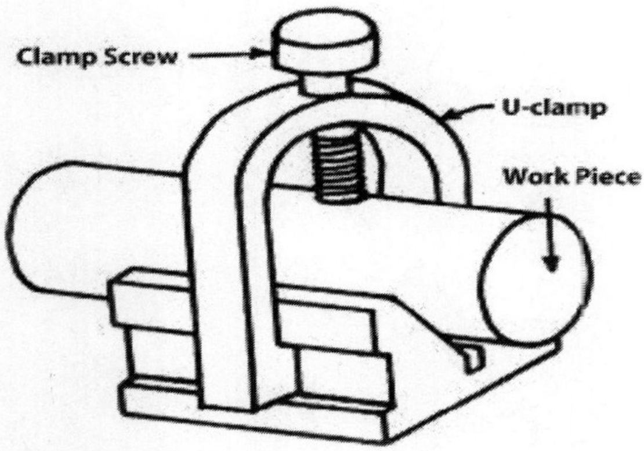


**Figure 94**. V Block

## 2. Cutting Tools

**i) Files:** These are multi-point cutting tools used to remove material by rubbing it on metals.

- Based on length: 4", 6", 8", 12"
- Based on a number of cuts:
- Single-cut files
- Double-cut files
- Rasp files

More material is removed in the case of double-cut files.

- Based on grade:
- Rough (R).
- Bastard (B).
- Second cut (Sc).
- Smooth file (S).
- Dead Smooth (DS).

Rough and Bastard are known as big-cut files. These files are used when the material is more. Dead smooth and smooth files have relatively smaller teeth. They are used for finishing work.

### Based of Shape and Size

The length of the file is between 4' to 14'. The various shapes available are:

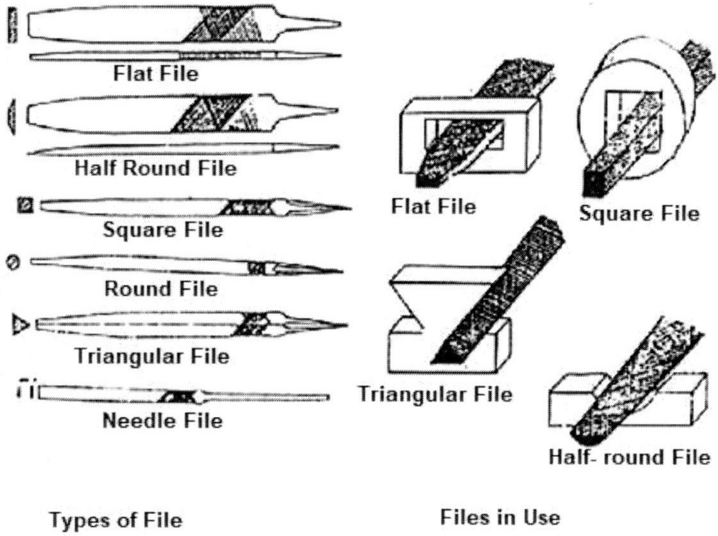


**Figure 95.** Types of Files

i) **Hacksaw:** The hacksaw is used to cut bars, rods, flats and other materials. It is made up of a frame made of mild steel and a fixed blade of high-carbon steel. Blades can be classified as forward and backward cuts according to their cutting directions. Blades are classified as Coarse, Medium, and Fine depending upon the pitch of teeth.

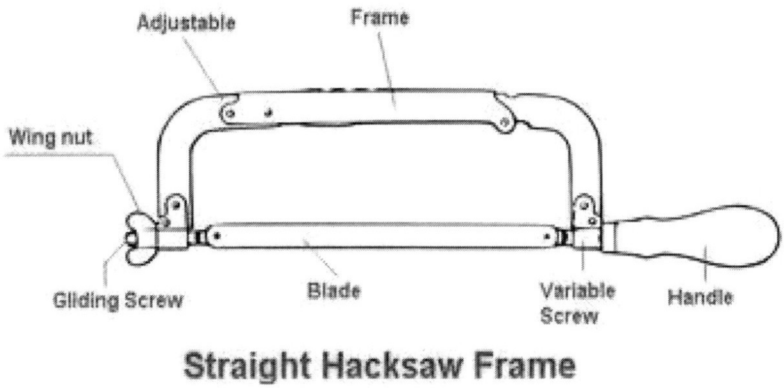


**Figure 96.** Hacksaw

ii) **Chisels:** Chisels are used for cutting metal and turning them into thin sheets and are used for removing excess metal. The cutting angle is about 60 degrees for a chisel.

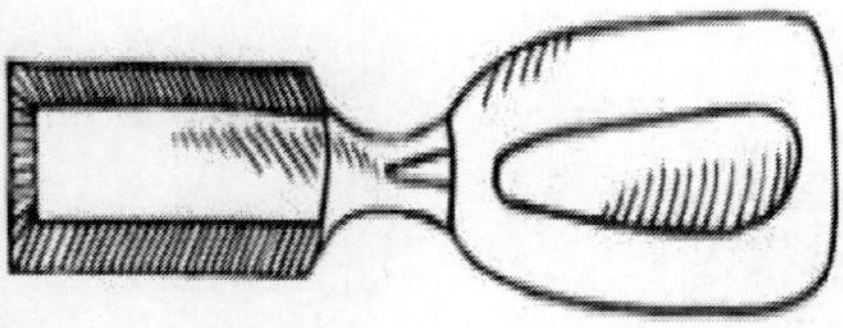

**Figure 97.** Chisel

## 3. Drilling Tools

i) **Twist Drill:** A twist drill is a rod made of metal with a specific diameter that consists of two, three, or four flutes with spiral shapes.

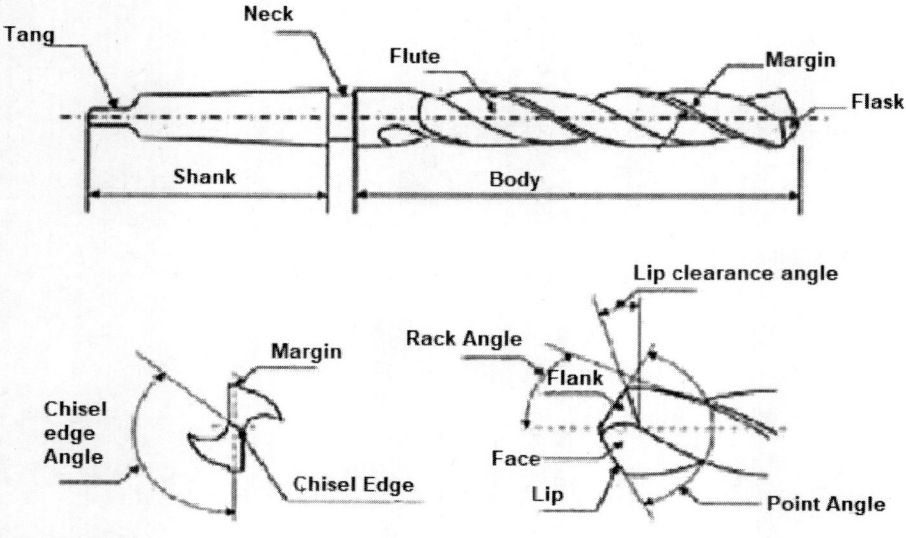


**Figure 98.** Twist Drill

ii) **Bench Driller Machine:** A bench driller machine can be, say, a small version of a pillar drill. A bench driller is used to drill lightweight objects or workpieces.

iii) **Reamers:** To provide finishing and to provide an accurate size to drill a hole with the help of a cutting tool called a reamer is known as reaming. It consists of a number of cutting edges.

For this process, we first have to drill a hole slightly smaller than the desired hole, and a reamer is used to provide a desired or finished hole.

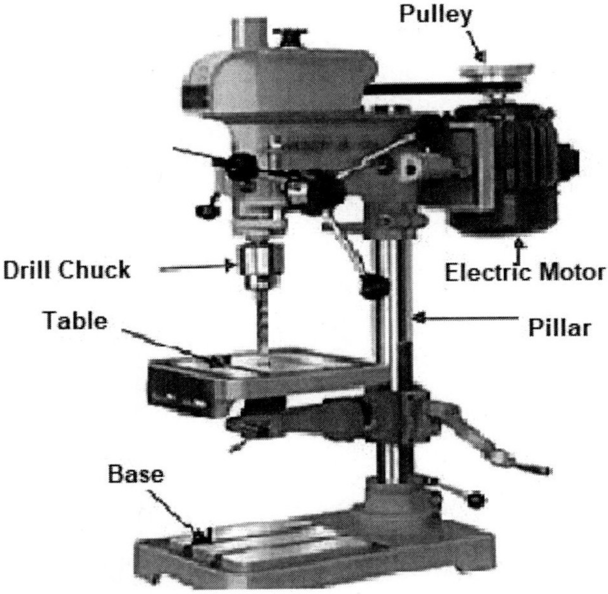


**Figure 99.** Pillar-Type Drilling Machine

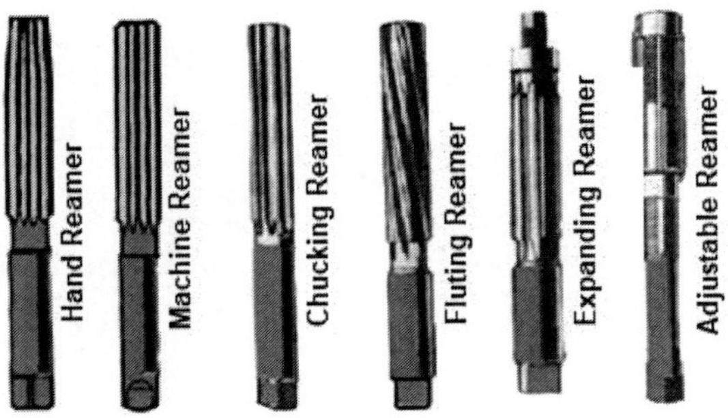


**Figure 100.** Types of Reamers

## 4. Threading Tools

i) **Tap Wrenches:** A tap wrench is a tool that is used to turn small taps. They are used to make screw threrds also.

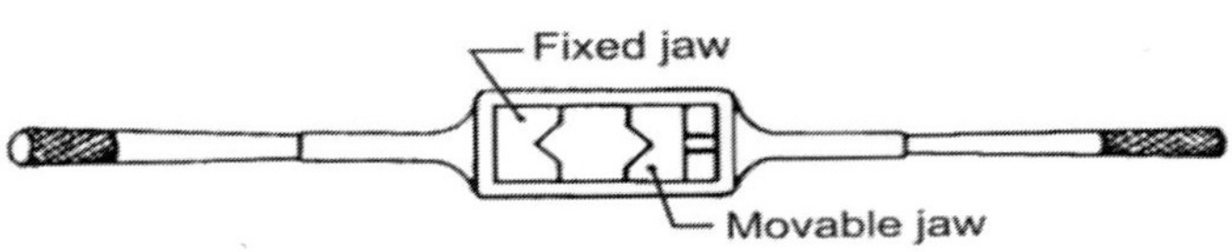


**Figure 101.** Tap Wrench

## 5. Measuring and Marking Tools

i) **Steel Rule:** The steel rule is made from stainless steel and has measurements from 12 feet to 2 feet and in inches and millimeters. Rough handling should be avoided near the edges of the steel rule.

ii) **Calipers:** There are various types of calipers to measure the inner diameters and outer diameters of objects. Outside calipers are used for measuring outer diameters, whereas inner calipers are used for measuring inner diameters. A spring caliper uses string to apply pressure. The lock nut is used to lock the desired position. Hermaphrodite, Odd leg, and Jenny calipers have one leg bent inward and the other end straight.

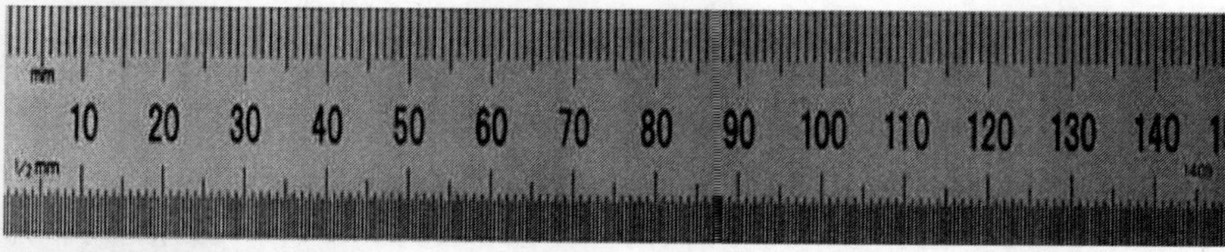


**Figure 102.** Steel Rule

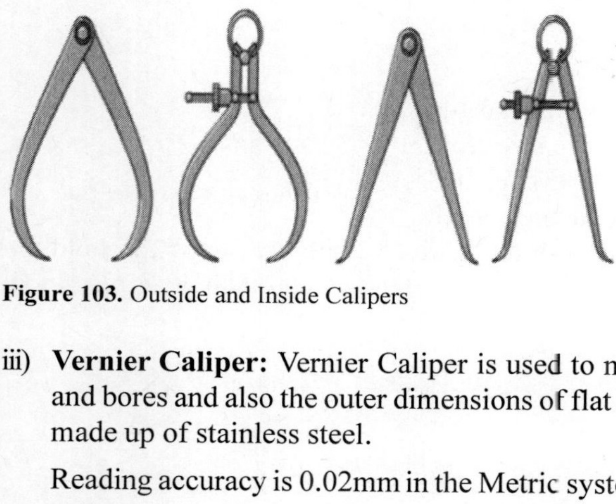

**Figure 103.** Outside and Inside Calipers

iii) **Vernier Caliper:** Vernier Caliper is used to make the inner size of holes and bores and also the outer dimensions of flat and round components. It is made up of stainless steel.

Reading accuracy is 0.02mm in the Metric system and 0.001" in the British system.

**Principle:** N divisions on the Vernier scale will be equal to the N-1 division on the main scale.

**Calculation for Least Count of Vernier Caliper:** The main divisions are millimeters.

Value on the main scale of each division = 1 mm= 1msd. Value of 1msd = 1 mm.

Value of vsd= (N-1)/N msd= 49/50 mm. Least count= 1msd – 1 vs = (1-49/50) mm= 0.02 mm.

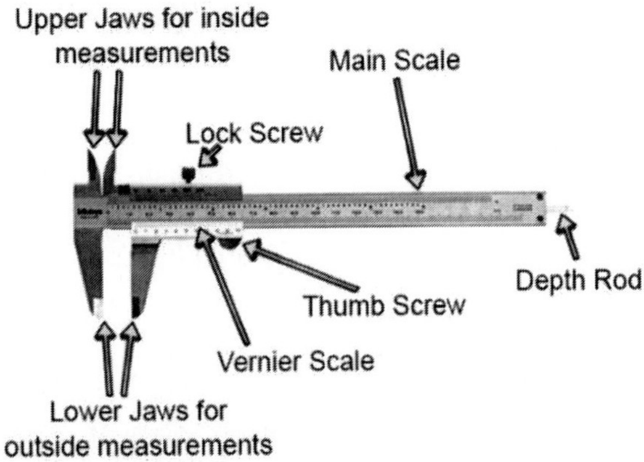


**Figure 104.** Vernier Caliper

iv) **Surface Plate:** It is used to test the flatness and trueness of the surfaces. It is made of graphite or cast iron. This tool is also used in Scribing work.

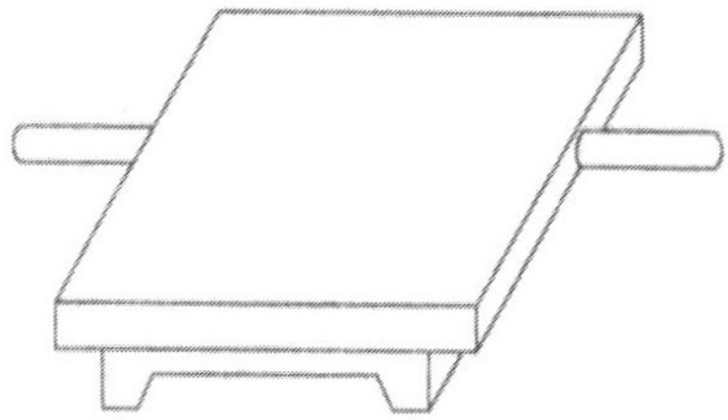

**Figure 105.** Surface Plate

v) **Angle Plate:** Angle plates are made up of cast iron and can range in size from small to large. They have two planed surfaces that are at right angles. Each surface has numerous slots through which bolts or clamps can hold the work. Hammering on the angle plate to tighten nuts and bolts is restricted.

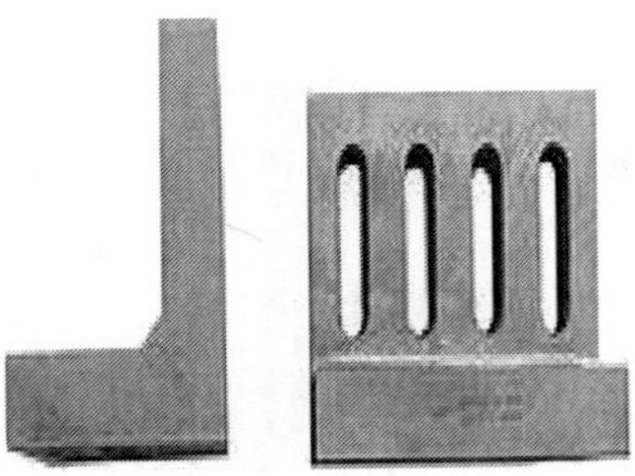

**Figure 106.** Angle Plate

vi) **Scriber and Surface Gauge:** A scriber is a marking tool used on metal workpieces. It is made up of slender steel (hardened & tempered carbon steel).

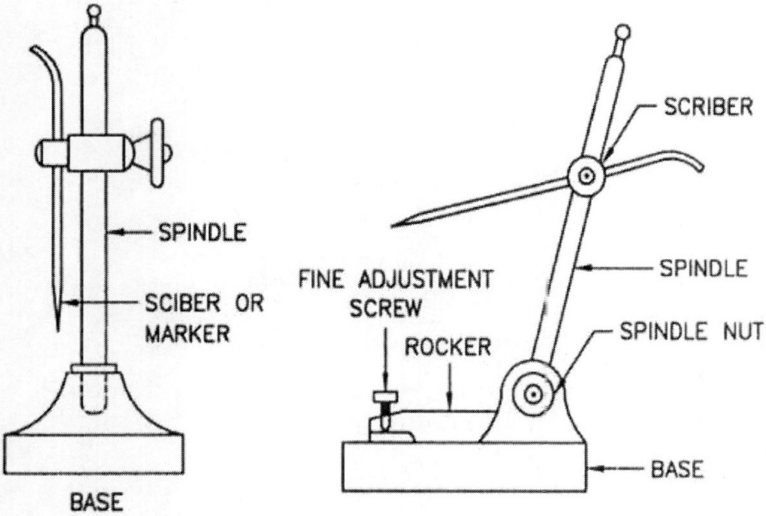


**Figure 107.** Scriber and Surface Gauge

vii) **Punches:** Punches are used for the purpose of marking. They are made of high-speed steel. Hammering is done while working on the second end.

- Dot punches.
- Centre punches.

**Figure 108.** Punche

viii) **Try-Square:** Try-square is used to check the squareness of two surfaces. It is also used for the making of right angles. Using Try Square as a hammer is restricted.

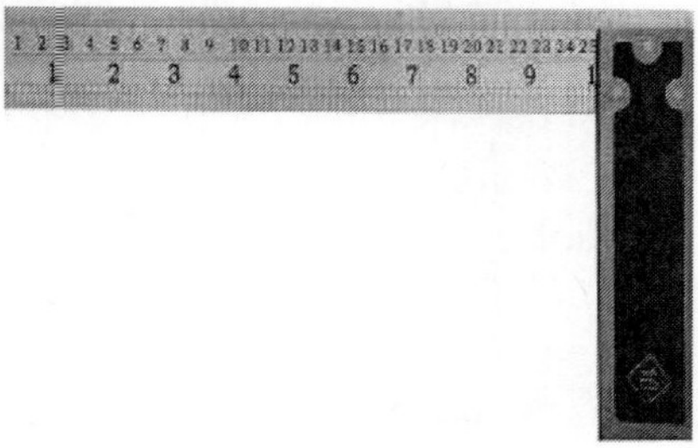

**Figure 109.** Try Square

ix) **Vernier Height Gauge:** It consists of four parts.

- A heavy base.
- A graduated beam.
- A sliding head with Vernier sliding.
- A fine adjustment clamp.

As opposed to Vernier Calipers, Vernier Height Gauge has a movable jaw, whereas Vernier Calipers have a fixed jaw.

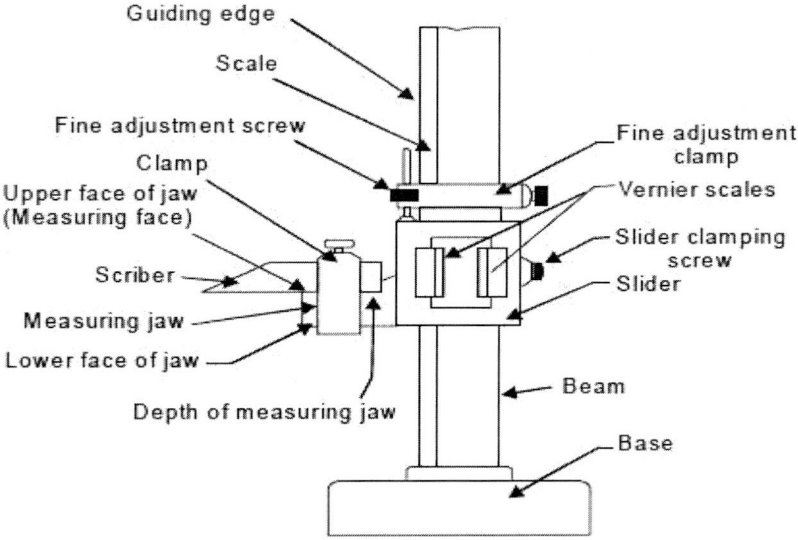


**Figure 110.** Vernier Height Gauge

## 6. Striking Tools

i) **Ball-Peen Hammer:** Ball-peen hammer is a hammer used for metalworking. It comprises two heads. One of them is flat, while the other is rounded called a peen. The use of this hammer is basically for striking and shaping metal materials. For hitting chisel and punches. A Ball-peen hammer can also be used.

**Figure 111.** Ball-Peen Hammer

ii) **Cross-Peen Hammer:** Cross-Peen hammer is a tool or hammer which is used to make the metal complete with the help of heat or when used with heat. A cross-peen has a blade that is perpendicular to the handle. It has sharp, round, smooth, flat, or maybe textured. It is also used for metalworking, woodworking, blacksmithing and so many other purposes.

**Figure 112.** Cross-Peen Hammer

iii) **Straight-Peen Hammer:** Straight-peen hammer is a hammer that is used by carpenters and has a peen parallel to the handle. Straight-peen hammer is similar to a Cross-peen hammer used for metal working, but in the straight- peen hammer, its peen is in line with the handle. It is also used for riveting in some areas or also for striking metals.

**Figure 113.** Straight-Peen Hammer

## Operations Used in Fitting Shop

A large number of operations are performed in the fitting shop to get the desired results.

1. Marking out
2. Sawing
3. Chipping
4. Filing
5. Scraping
6. Grinding
7. Drilling
8. Reaming
9. Tapping

1. **Marking Out:** Marking is the operation how to get an outline of the job to be performed it is done by marking specific lines and positions. Marking gives the overview to perform fitting operations to get a desired shape and size workpiece. Call markings are done with respect to the pre-decided reference. We commonly make use of chalks or copper sulfate solution to give markings.

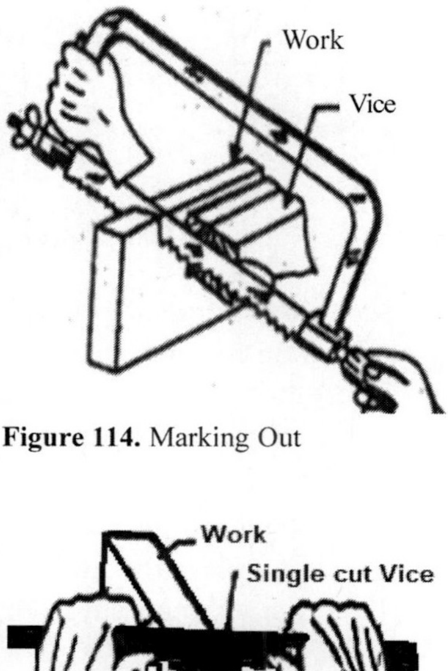

**Figure 114.** Marking Out

2. **Sawing:** Sawing is the operation used to cut the workpiece apart or give it the desired shape or make slots in it. In this operation, the workpiece is fixed on the vice and the Blade is fixed with the help of screws and then, using both our hands, we move the blade forward and backward and do the desired cutting or shaping.

**Figure 115.** Sawing

3. **Chipping:** Chipping is the operation used when we need to clear out the thick layers of metal from the surface of the workpiece using a chisel. For chipping operation, the workpiece is fitted tightly in the voice and then the metal is removed by hitting the chisel on its back with a hammer.

4. **Filing:** The filing operation is used to give the workpiece better finishing by removing the access materials present on the workpiece after chipping. It is performed using a file. There are different types of files available for different types of workpieces and required results. It is used to give the workpiece are smooth finish. We need to move the file on the workpiece fitted tightly on a wise and decide movement while stressing on the forward stroke and releasing the pressure on the backward stroke.

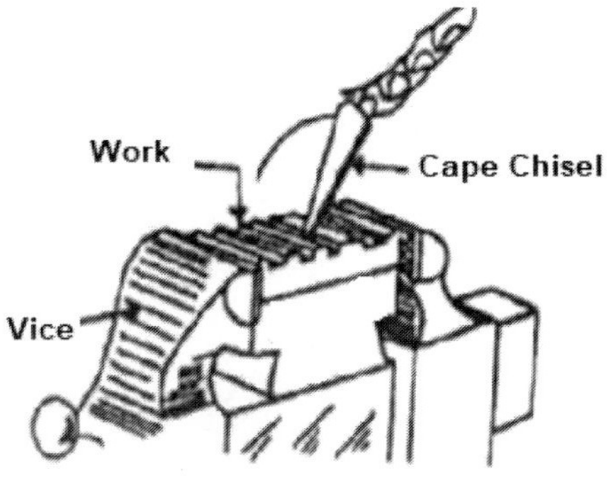


**Figure 116.** Chipping

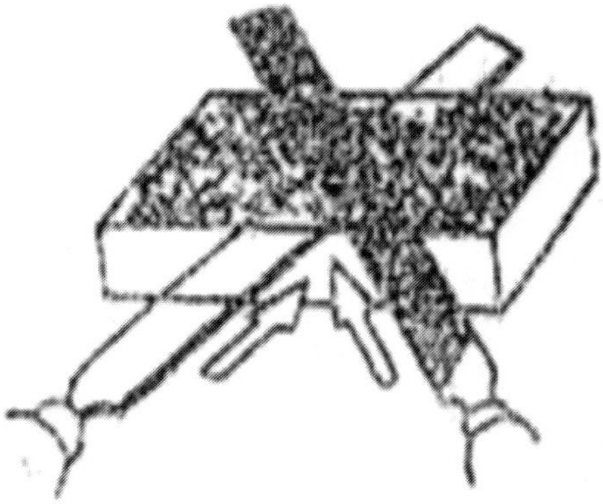

**Figure 117.** Filing

5. **Scraping:** Scrapping is an operation to be performed by human hands and is only used when a machine cannot do the required finishing of the workpiece. Types of scrapers are used to smoothen the surface of a workpiece and clear the defects after major operations are performed on it.

6. **Grinding:** Grinding is the operation used to remove the material from the required part of the workpiece using a grinding wheel. The grinding wheel is rotated against the surface of a workpiece and during its motion, it removes the unwanted material from the workpiece.

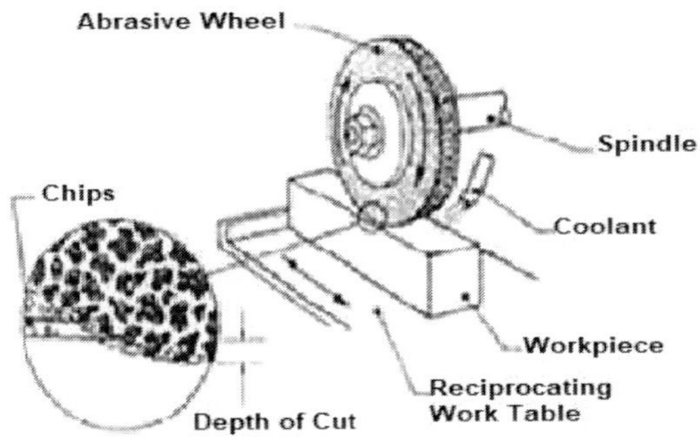


**Figure 118.** Scraping

**Figure 119.** Grinding

7. **Drilling**: Drilling is the operation used to produce a hole in the workpiece or cut a circular cross-section from the workpiece. In drilling, a long rotating rod equipped with a tool is used to make a hole or cut the desired cross-section.

8. **Reaming:** Reaming operation is performed post-drilling to give a better-required finish to the already drilled hole. The shape of a hole after the drilling operation is not geometrically e accurate and smooth hence reaming comes in here. It is performed using a multipoint rotating tool called "Reamer". To the back of the machine and then the feed is given to the workpiece fitted in a vice.

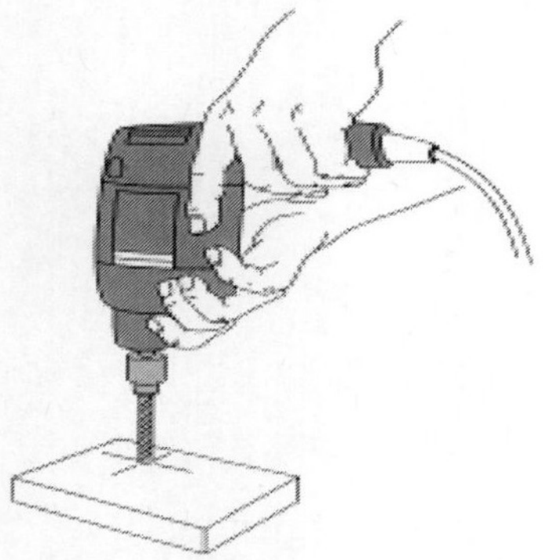

**Figure 120.** Drilling

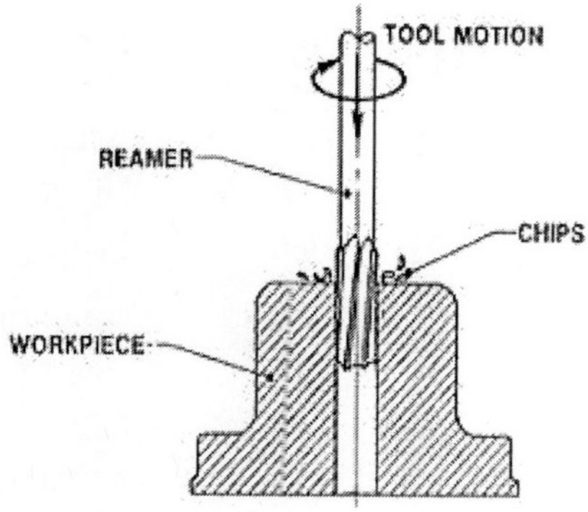


**Figure 121.** Reaming

9. **Tapping:** Tapping is the operation performed to cut internal threads in the hole to make it feasible for the bolt or cap screw to fit into the hole. It is performed by an instrument called a tap. Taps are of different types depending on the size of the teeth they carry to make the required size of internal threads.

**Figure 122.** Tapping

## Experiment 1

**Aim:** To make mild steel plate into the desired or required model with the help ofT-fitting.

### Materials Required

- Mild steel plate of dimensions 48 mm in length
- 34 mm breadth.

### Tools Required

- Bench Vice.
- Punches.
- Scriber.
- Hacksaw with its blades.
- Files.
- Flat chisel.
- Steel rule.
- Ball-peen Hammer.
- Files.
- Try square.
- Surface plate.
- Vernier height gauge.

### Theory

Following are the series of operations carried out in this experiment: -

1. Filing (metal removal process).
2. Checking the flatness and squareness.
3. Marking & Measuring.
4. Punching- Forming process (through the workpiece) to create a hole.
5. Sawing- It is for separating parts into two or more pieces.
6. Chipping- To remove excess metal with cold chisels.
7. Finishing- It provides smoothness or touch to the workpiece through this method.

### Procedure

1. Take two pieces of mild steel having dimensions of 48x 34 mm.
2. Then hold the workpiece, and check the alignment of the surface with the help of Try-square.
3. The surface should be mutually perpendicular to the other surface.
4. Then hold the Try- square and workpiece together and mark on the surface with the help of the scriber.

5. An excess amount of material is removed by the filing.
6. After this workpiece is held on the Bench Vice and unlocked, the liver and tightens to hold the workpiece between the jaws of a bench vice.
7. Now with the help of the flag file start filing the surface.
8. After filing, the workpiece is taken to the Vernier height gauge.
9. Now chalk is rubbed to see the marking by Vernier height gauge.
10. After this, hold the workpiece against the angle block and mark it as per the sketch.
11. Now using the dot punch and ball-peen hammer, put the workpiece on the angle.
12. Small marks are made along the lines so that cutting can be done easily.
13. Hold the workpiece between the jaws of the bench vice perfectly so that cutting can be done straight.
14. After cutting on one side, the other side is also cut in the same way.
15. Similarly, cutting is done by a hacksaw on the second workpiece.
16. By holding the workpiece between the jaws of a bench vice.
17. The surface becomes filed and makes it plane.
18. Both surfaces are filed.
19. Similarly, the second workpiece is done.
20. Now after filing, the jaws of the bench vice are unlocked, and taking the two workpieces, their alignment is checked.
21. The perfect alignment is acquired.

**Result:** The required square T-Fit is made as per the dimensions taken.

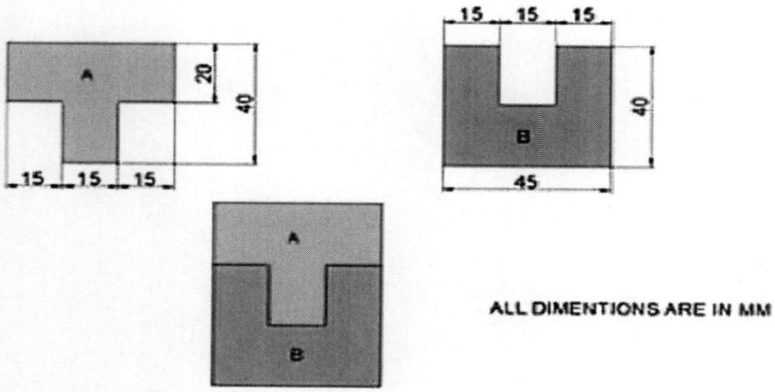


**Figure 123.** Mild Steel Plate Model

## Precautions

1. Remember marking dots should not be crossed.
2. Do not apply pressure in the backward direction; hence it should be in the forward direction.
3. Do not rub the steel rule on the job.
4. Before use, files should be cleaned carefully.
5. Blades should be fixed with proper tension in hack sawing.
6. Precision instruments like Vernier height gauges are carefully used.
7. Lubricant oil should be used during hack sawing.

## Experiment 2

**Aim:** To make a mild steel plate into the desired model or piece with the help of V-fitting.

**Materials Required:** Mild steel plate of dimensions 48 x 34 mm.

### Tools Required

- Bench Vice.
- Try square.
- Hack saw and blade.
- Dots & Centre punch.
- Surface plate.
- Rough & Smooth flat files.
- Steel rule.
- Ball- peen hammer.
- Scriber.
- Vernier height gauge.
- Flat & Triangular chisel.

### Theory

Following are the series of operations carried out in this experiment: -

1. Filing (Material removal process).
2. Checking flatness and squareness.
3. Marking and Measuring.
4. Punching: Forming process (through the workpiece) to create a hole.
5. Sawing: It is for separating parts into two or more pieces.
6. Chipping: To remove excess metal with cold chisels.
7. Finishing: It provides smoothness or touch to the workpiece through this method.

### Procedure

1. Take two workpieces made up of mild steel of 48x34 mm.
2. Then hold the workpiece with the help of Try-Square.
3. Then check the alignment of the surface such that the surface is mutually perpendicular to the other surface.
4. Now by holding the workpiece and Try-Square together.
5. Take the scriber and mark the surface such that excess metal should be removed by filing.
6. After that, the workpiece is held on the bench vice.

7. Unlock the lever and tighten it to hold the workpiece between the jaws of a bench vice.
8. Now start filling the surface.
9. After filing the workpiece is then taken to the Vernier height gauge.
10. Chalk is rubbed against the workpiece to see marking that will be done by Vernier height gauge.
11. Holding the workpiece against the angle block marking is done as per the sketch.
12. Three lines are made of the desired dimensions.
13. After doing the marking, the edges are joined by a scriber with the help of a steel ruler to give the shape of a V.
14. With the help of a dot punch (and/or ball-peen hammer) put the workpiece on the angle.
22. Small marks are made along the lines so that cutting can be done easily.
23. Hold the workpiece between the jaws of the bench vice at an angle that cutting can be done straight.
24. Cutting is done by hacksaw on a first workpiece.
25. After doing the cutting, the jaws of the bench vice are unlocked.
26. In this way, the required dimension of V cutting is obtained as per the sketch.
27. Similarly, cutting is done by a hack saw on the second workpiece.
28. Now in this way required dimension of V cutting is obtained as per the sketch.
29. Finally, surfaces are filed to make it plane.
30. Both surfaces are filed.
31. The jaws of the bench vice are unlocked and the second workpiece is put in between the jaws.
32. The filing is done on the first side.
33. Turn the other side and continue filing.
34. After filing the jaws of the bench vice are unlocked.
35. Taking the two workpieces, the alignments are checked.
36. The perfect alignment is acquired.

**Result:** The required V- fitting is obtained with the mentioned dimensions.

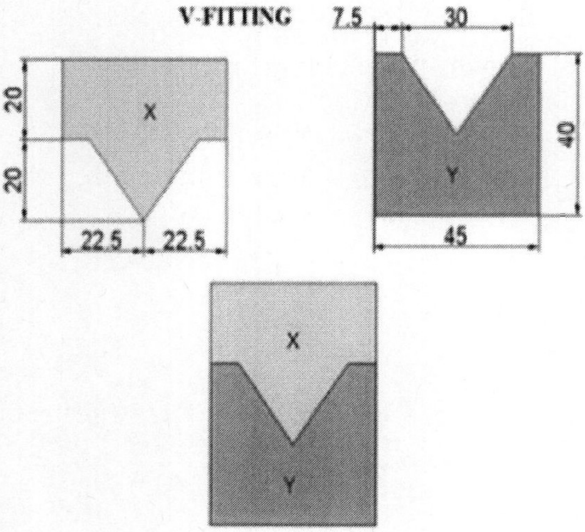


**Figure 124.** V Fitting

## Precautions

1. While marking, we should take care of marking.
2. During hack sawing, we must take care.
3. While removing chips brush should be clean.

## Experiment 3

**Aim:** To understand the method of Marking in the Fitting Shop.

### Tools Required

- Surface Plate.
- Scriber and surface gauge.
- Angle Plate.
- Punches.
- Try-Square.
- Vernier Height Gauge.
- Try Square.

### Procedure

1. The work should be held with the help of a holding device according to its shape and size. For a flat workpiece, use a surface plate.
2. Horizontal direction lines are scribed by surface gauge.
3. After scribing the workpiece, a dot punch and hammer are used on the surface to make indentations.

### Result

The V-fitting of the required dimensions is obtained.

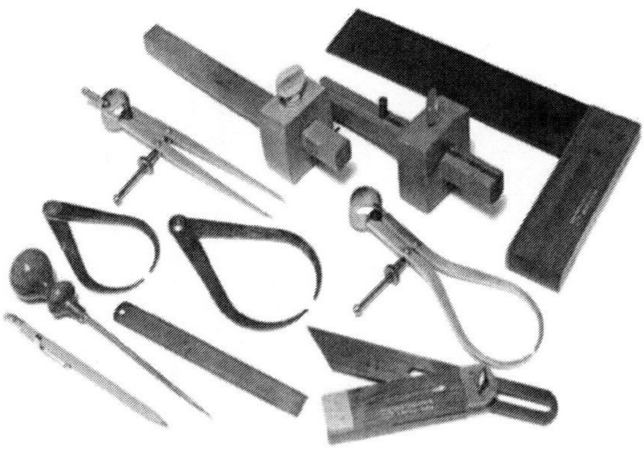

**Figure 125.** Tools

### Precautions

1. Care should be taken by hacksawing.
2. Marking should be done carefully.

## Experiment 4

**Aim:** To make a fitting job as per drawing.

### Tools Required

- Try square.
- 200 gm hammer.
- Dot punch.
- Steel rule.
- Scriber.
- Odd leg caliper.
- Hacksaw.
- Shank drill.
- Rough flat files.
- Half-round files.

### Theory

Drawing in the fitting shop is needed for prefabricated components. Shop drawing is the drawn version of information (stored in construction documents). Drawing is done to emphasize the fabrication and installation of the items. The style of shop drawing is quite different from the drawing of an architect.

### Procedure

1. The given material is to be checked for 48x48x5 mm dimensions.
2. All the edges are checked for straightness.
3. Filing is done to make 46x46 and a thickness of 5mm.
4. Note the portion which is to be removed and mark it.
5. Using Dot-punch, punching is done along the lines.
6. Excess material is cut with the Hacksaw to make it a correct fitting.
7. For the inside radius, excess material is removed by Hacksaw and then filing is done by flat rough file.
8. For the outside radius, excess material is removed by Hacksaw and then filing is done by flat rough file.
9. Using a drill bit and bench drilling machine are drilled as per the drawing.
10. Final blurs are removed by the filling surface of the fitting job.

### Scope of Resolution

1. While working, we must goggle ourselves, so that dust does not enter our eyes.
2. Wear a glove while using a hacksaw to prevent your hands from cutting.
3. We must have knowledge of various machines in the workshop.

**Result:** The Fitting job is obtained of required dimensions.

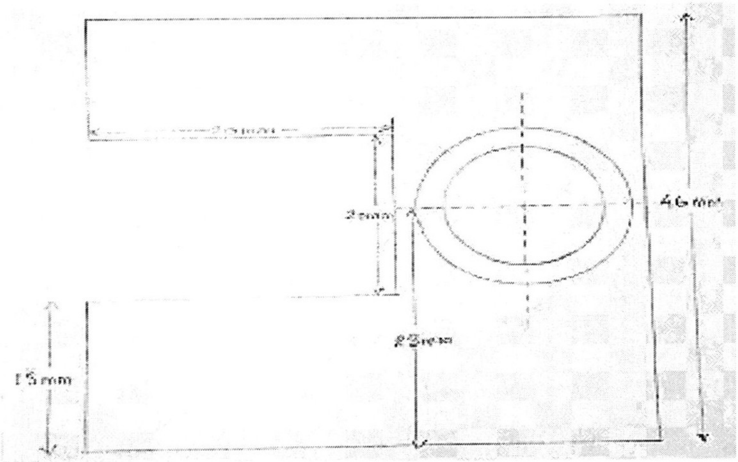


**Figure 126.** Fitting Job Drawing

## Precautions

1. Keep proper discipline in the shop.
2. Do not play with the tools.
3. While using a Hacksaw, make sure its tooth is sharp and in the forward position.
4. While using a Chisel, the direction should be away from the working person.

## Fitting Shop Objective Questions

1. Fitting vice is a ——————.

   **a) Holding Vice for Metals** b) Holding Vice for Woods

   c) Cutting Device d) Measuring Device

2. Chisel is a————————————.

   **a) Cutting Tool** a) Holding Vice for Metals

   c) Measuring Tool d) Striking Tool

3. Try square is a ——————————.

   a) Marking Tool b) Measuring Tool

   **c) Both 1 and 2** d) Cutting Tool

4. Tap wrench is a———————.

   a) Cutting Tool b) Marking Tool

   **c) Holding Vice** d) Measuring Vice

5. Which of the following instruments is used to measure the length and width of the job piece?

   a) Try Square **b) Oddleg Caliper and Scale**

   c) Outside Caliper and Protector d) None of these

6. Which of the following has one movable jaw and one fixed jaw?

   a) Hammer b) Mallet

   **c) Tap Wrench** d) None of these

7. Centre punch is a ————- .

   a) Holding Tool b) Striking Tool

   **c) Cutting Tool** d) MarkingTool

8. By which instrument can you mark and draw the vertical lines on the job piece?

   a) Oddleg Caliper b) Centre Punch

   c) Scriber **d) Try square**

9. Ball Peen hammer is a ——————.

   **a) Striking Tool** b) Cutting Tool

   c) Measuring Tool d) Marking Tool

10. Which of the following is not a cutting tool.

    a) **Try Square** b) File

    c) Rip Saw d) Chisel

## Fitting Shop Subjective Questions

1. How a hammer is specified? What is the main difference between the hand hammer used in a smithy shop and a fitting shop?
2. State the difference between a file and a scrapper in regard to their use.
3. How many types of punches are used in bench work?
4. Write short notes on the following .

    a) Chipping b) Grinding

    c) Drilling d) Tapping

# 3

# Sheet Metal Shop

## Introduction

Figure 127. Sheet Metal

A sheet metal shop is a kind of workshop where different types of materials are used to make sheet metal and then it is flattened into thin sheets of desired shape and size. Sheet metals are used for the production of automobile bodies, utensils, almirah and cabinet appliances, etc.

Sheet metal parts have many advantages, like they are less expensive, lighter in weight and can be modified into versatile shapes. Low-carbon steel is commonly used. The use of casting and machining is being replaced by metal stamping. The stamping press is a kind of machine tool in which most of the operations occur, and the sheet metal products which are obtained by the machine are called stampings.

## Sheet Metal Work

Sheet metal is a plate having a thickness of about less than 5 mm. Products made by sheet metal processing are almirah, electrical parts, utensils & buckets etc. In some cases, the products of sheet metal are used to replace the use of forging and casting.

Sheet metal produces many products, which fulfill the need of households, and decoration works. Further machining is not needed in sheet metal working, as required in casting and molding. The time taken by sheet metal working is approximately half as compared to the time taken in the machining process. For doing sheet metal work, geometrical knowledge and properties of a metal are needed for better results.

## Metals Used in Sheet Metal Work

1. Black Iron Sheet

   This metal has a very low cost and also can be rolled easily.

i) If it is uncoated, then it corrodes, so for using it for a long period, it must be painted or enameled.

ii) Tanks, food containers, furnace fittings and dairy pieces of equipment are some examples of black iron sheets.

2. Galvanized Iron (also known as G.I.)

i) Soft steel and a layer of molten zinc are coated over it.

ii) Buckets, mugs, and cabinets are some examples of G.I. metal sheets.

3. Stainless Steel

i) The alloy of steel with nickel, chromium and some other metals is very costly.

ii) Kitchen wares, food handling equipment and tools and instruments which are used in hospitals are some examples of stainless steel.

## Tools Required in Sheet Metal Work

### 1. For Measuring and Marking

Measuring tools are used to measure the metal sheet to make it easy to work on the sheet properly and accurately.

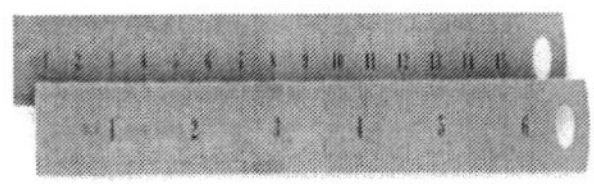

**Figure 128.** Steel Rule

**Tools**

i) **Steel Rule:** Steel rule is used for linear measurement.

**Figure 129.** Scribers

ii) **Scribers:** It is a tool made up of hardened steel. It is used for pointing lines.

iii) **Divider:** It is used for drawing circles and arcs.

iv) **Swing Blade Protractor:** The swing blade is used for marking and measuring angles.

v) **Wire Gauge:** A wire gauge is used to measure the thickness of the sheet.

**Figure 130.** Divider

**Figure 131.** Swing Blade Protractor

**Figure 132.** Wire Gauge

## 2. Striking Tools (Hammers and Punches):

i) **Hammers:** In engineering, it is a hand tool that is used for strengthening and stretching.

- **Smoothing Hammer**: used for making the sheet metal joint smoother.
- **Creasing Hammer**: used to close down the edges of the joints of sheet metal parts.
- **Stretching Hammer**: used for stretching the sheet.
- **Hollowing Hammer**: used for making the hollow part in the sheet metal.
- **Riveting Hammer**: rivet heads are formed by a riveting hammer.
- **Planishing Hammer**: used for removal of small marks from the sheet metal surface.

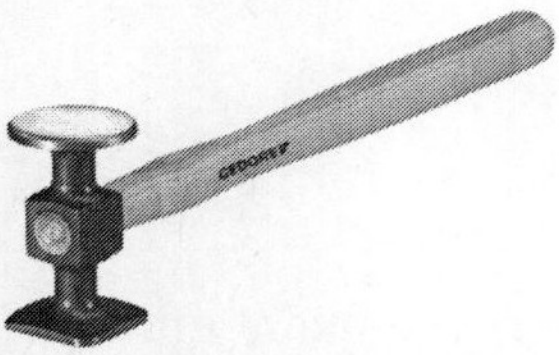

**Figure 133.** Smooth hammer

**Figure 134.** Creasing hammer

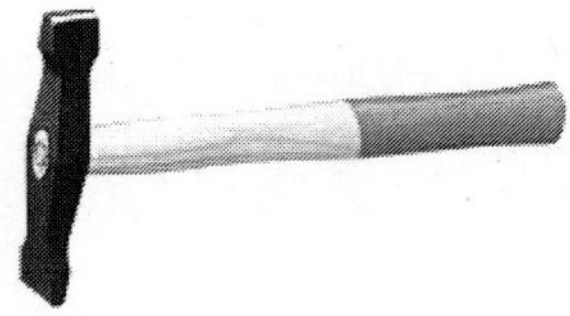

**Figure 135.** Stretching hammer

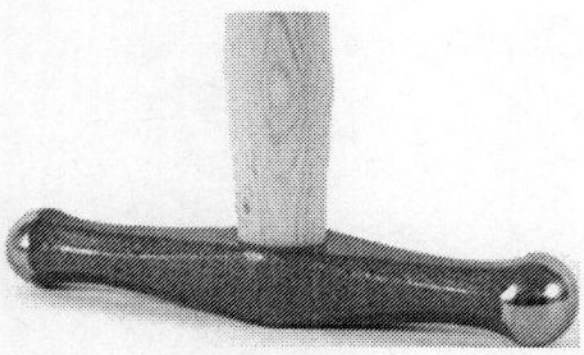

**Figure 136.** Hollowing hammer

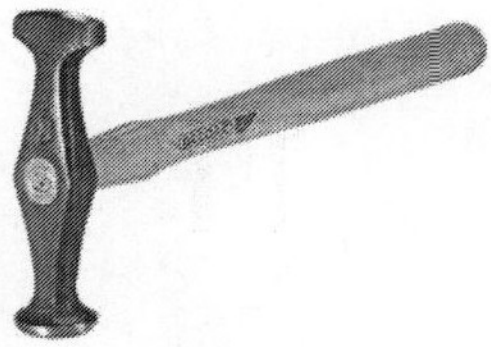

**Figure 137.** Planishing hammer

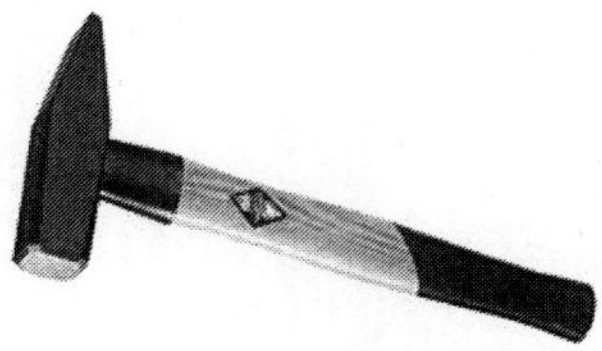

**Figure 138.** Riveting hammer

ii) **Punch:** For locating the center in metal sheets, punches are used. It generally has four types.

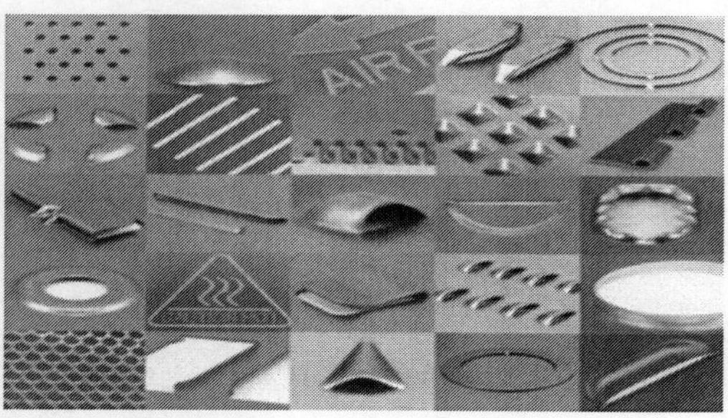

**Figure 139.** Punch

## 3. Cutting Tools (Chisels, Snips or Shears):

i) **Chisels:** Chisels are used to cut the undesired metal from the surface of a job by chipping. Flat chisels are used for making flat surfaces and it cut at an angle of 60 degrees.

ii) **Snips or Shears:** Snips or Shears are like scissors that are used for cutting thin metal sheets. Its length varies from 200 mm to 600 mm.

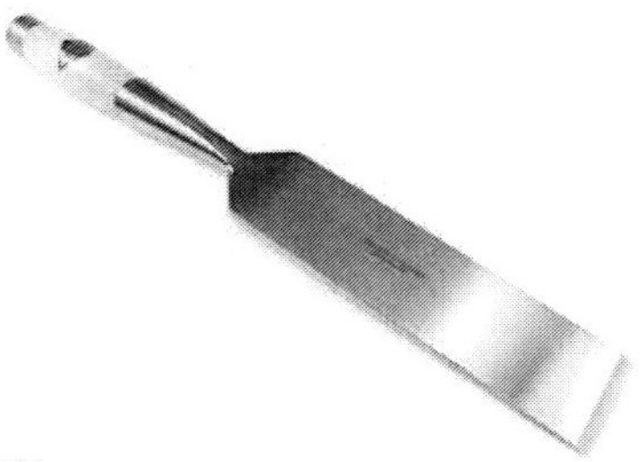

**Figure 140.** Chisels

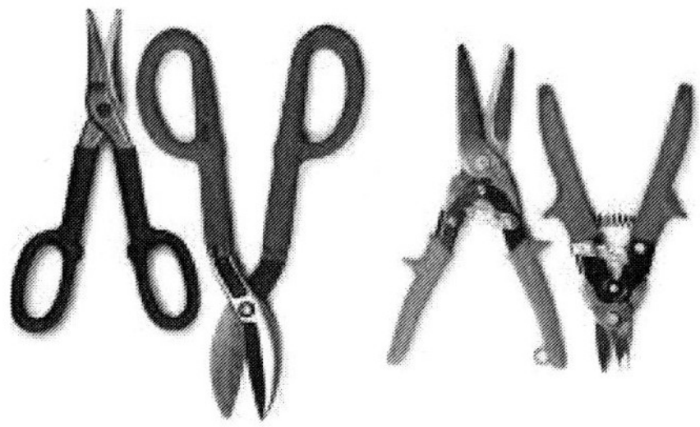

**Figure 141.** Snips or Shears

## 4. Bending Tools (Pliers, Folding Bars):

i) **Pliers:** In metal sheets for bending operation, pliers are mainly used to bend the sheet metal as per our requirement. Pliers are used for holding and cutting.

ii) **Folding Bars:** They are used for bending and folding operations.

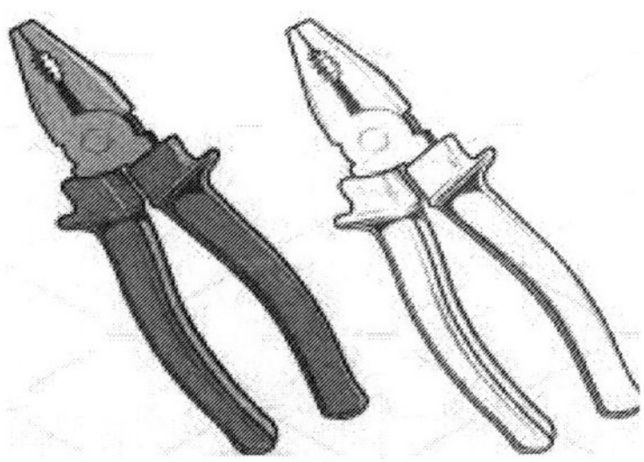

**Figure 142.** Pliers

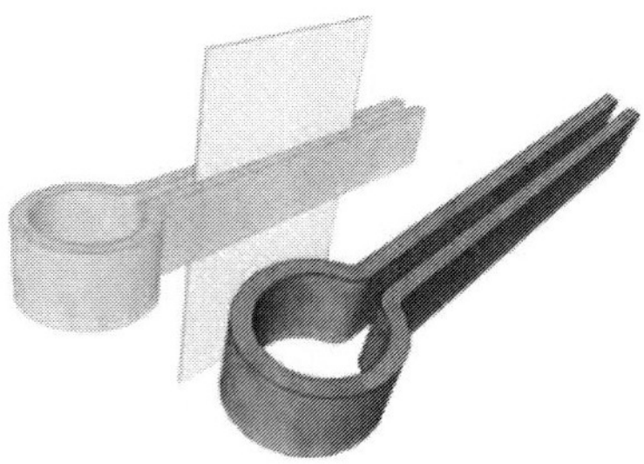

**Figure 143.** Folding bars

## 5. Supporting Tools (Stakes, Anvils, Swage Block)

i) **Stakes:** These are used for forming, bending seaming, etc. There are many shapes of stakes are available, which are based on the purpose for which they are used.

ii) **Anvil:** It acts as a base for smithy work when we do hammering. The weight of the anvil is up to 100 to 150 kg.

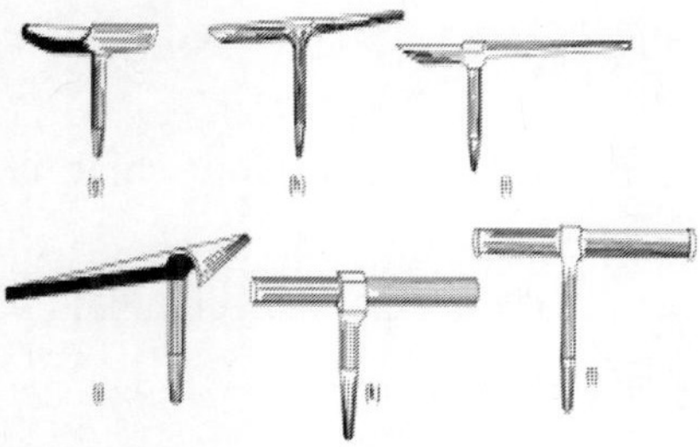

Figure 144. Stakes

Figure 145. Anvil

## Operations Used in Sheet Metal Shop

1. **Shearing:** Shearing takes place when the punch strikes and forms a cut in the sheet, which is placed on the die. In between the shearing edges of the punches and dies, the quality of the cut surface is great and clear.

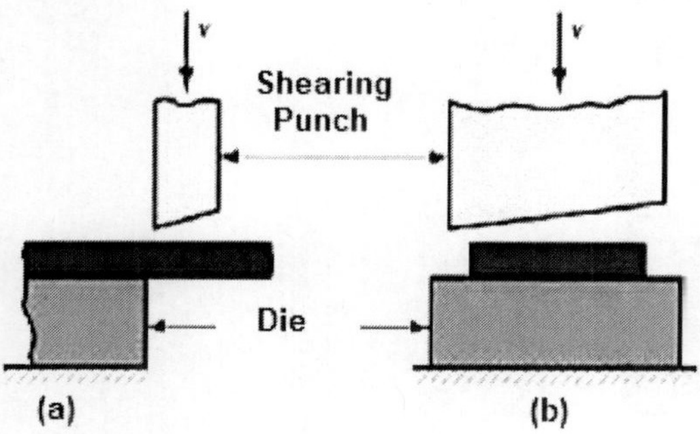


**Figure 146.** Shearing

2. **Cutting:** It is a process in which a piece from a sheet or strip can be cut/ separated in a line using the punch and die or press machine.

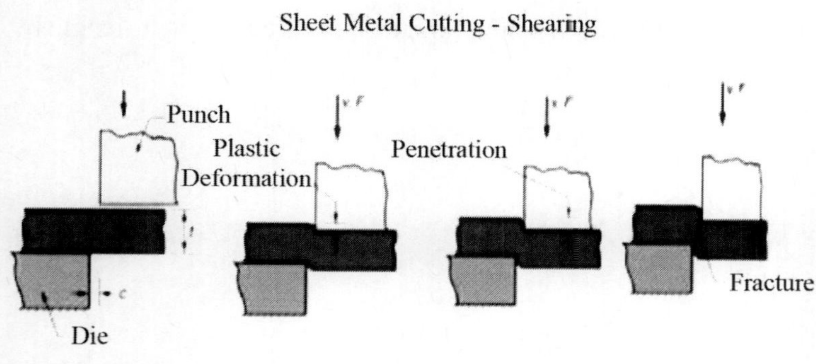


**Figure 147.** Cutting

3. **Parting:** In this process, between two pieces, scrap is removed to part them using a suitable press machine.
4. **Blanking:** This process involves the removal of some portion which is called a blank.
5. **Punching:** In this operation, circular holes are produced in sheet metal by using a punch and die of the press tool.

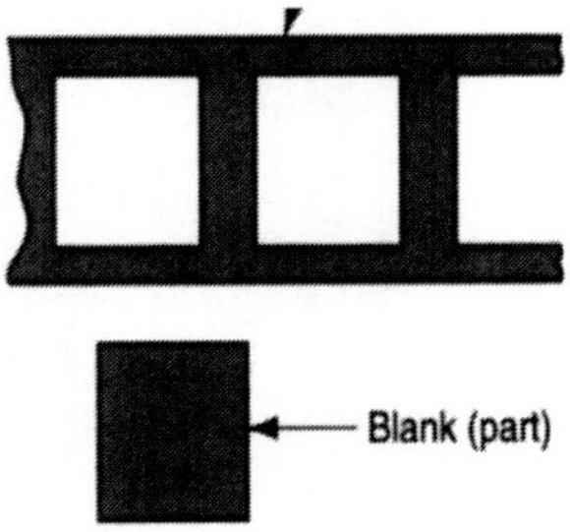


Figure 148. Blanking

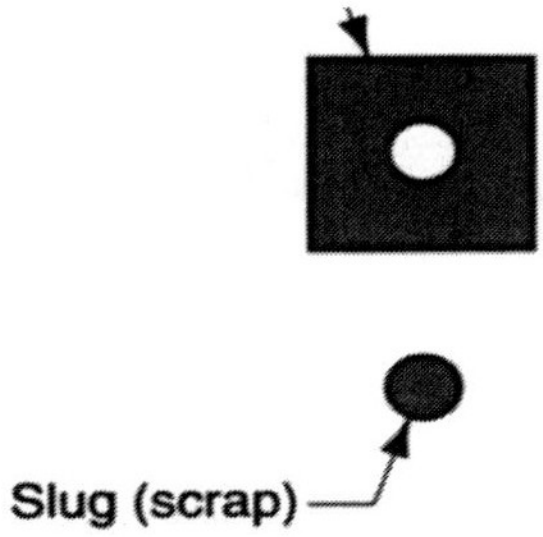


Figure 149. Punching

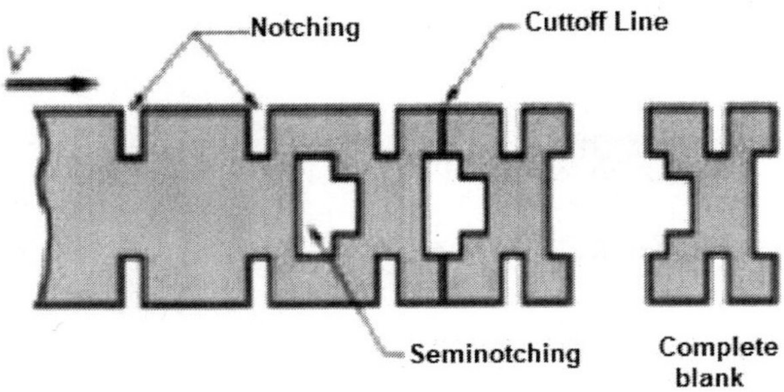


**Figure 150.** Notching

6. **Notching:** A specified shape of metal is cut from the side or the stock using a press machine.
7. **Slitting:** Slitting is a process in which a metal sheet is cut along its length by using a slitting machine.

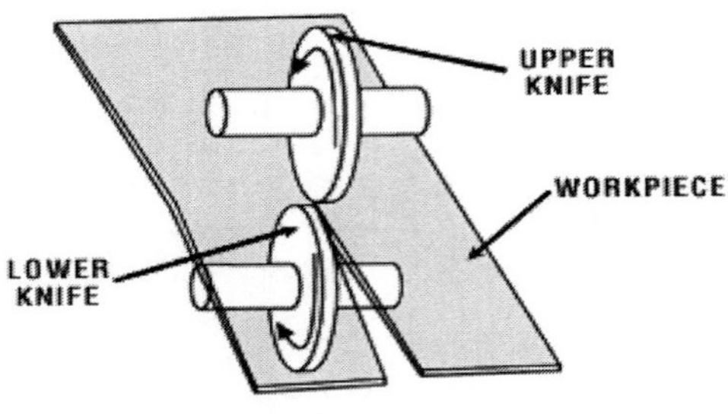


**Figure 151.** Slitting Process

8. **Nibbling:** In this cutting operation desired shape from sheet metal, without using a special tool, is obtained.
9. **Trimming:** Excess metal is cut out in flange or flash from a sheet metal part.
10. **Bending:** Deforming the sheet around a straight axis is done in this operation.

**Figure 152.** Nibbling

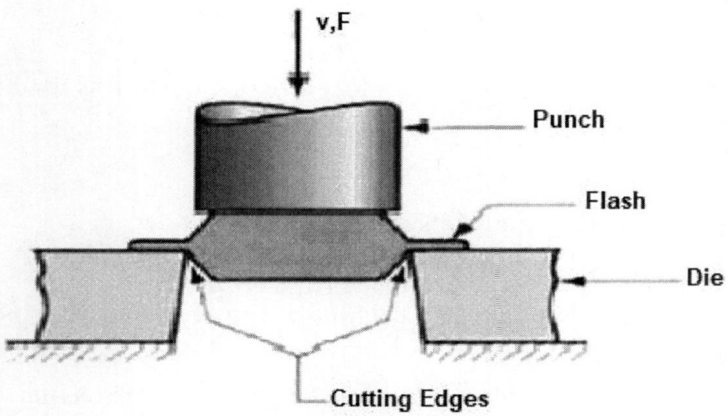


**Figure 153.** Trimming

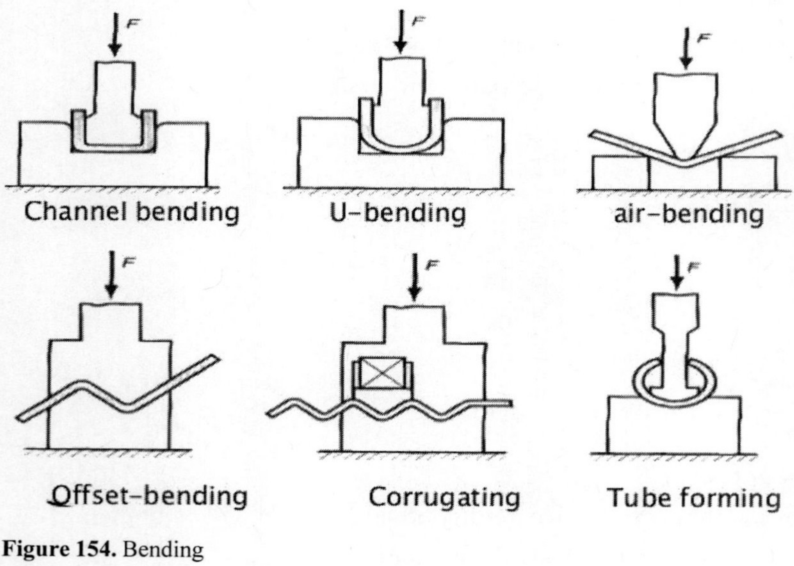


**Figure 154.** Bending

## Experiment 1

**Aim:** To make a Riveted Lap Joint.

**Materials Required:** Rivets, G.I. sheet.

### Tools Required

- Steel rule.
- Try square.
- Shear.
- Hammer.
- Mallet.

### Theory

**Lap Joints:** A joint in which one sheet is joined over the other. It is used in sheet metal work.

### Procedure

1. Take the measurement for cutting the sheet; use a steel ruler.
2. Check the straightness of the sheet.
3. Cut the G.I. sheet strip into two pieces with the help of a straight snip of suitable dimensions.
4. Now remove the sharpened edges and mark the points where holes are to be made to put the rivets inside it.
5. The distance between the holes of the rivets should be equal.
6. Make the holes on the sheet.
7. Put the rivet in the holes and apply hammering on it.

   **Result:** Riveted lap joint is prepared successfully.

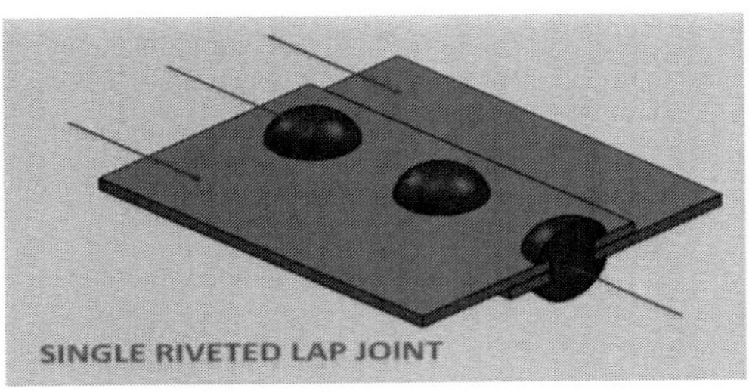


**Figure 155.** Single Riveted Lap Joint

### Safety Precaution

1. Hammering should be done carefully otherwise rivet will be broken.
2. Sharp tools must be used to avoid any injury.

## Experiment 2

**Aim:** To make a simple Hem Joint.

**Materials Required:** G.I. sheet of suitable dimensions.

### Tools Required

- Steel rule.
- Shear.
- Mallet (wooden hammer).
- Groover.
- Try square.
- Hammer.
- Straight snip.

### Theory

**Hem Joints:** Sheet metal edges are enabled by this joint for joining the pieces along with them.

### Procedure

1. Take the measurement for cutting the sheet using a steel ruler.
2. Check the straightness of the sheet.
3. Cut the G.I. sheet strip with the help of a straight snip of suitable dimensions.
4. Now remove the sharpened edge and fold the edge of the sheet once by using a mallet (wooden hammer) and groover.

   **Result:** Hem Joint is made as per the requirements.

**Figure 156.** Hem Joint

### Safety Precautions

1. After the completion of work, keep the tools in the proper place.
2. Care must be taken at the time of cutting of sheet.
3. Sharp tools must be used to avoid any injury.
4. A supporting tool must be taken while hammering.

## Experiment 3

**Aim:** To Make a Simple Seam Joint.

**Materials Required:** G.I. sheet of suitable dimension.

### Tools Required

- Scriber, steel rule.
- Try square.
- Shear.
- Hammer.
- Mallet (wooden hammer).
- Straight snip.
- Groover.

### Theory

**Simple Seam Joint:** It is a locked joint that is used to ensure a very positive group.

### Procedure

1. Take the measurement for cutting the sheet using a steel ruler.
2. Check the straightness of the sheet.
3. Mark the marks on the sheet using the scriber.
4. Cut the G.I. sheet strip with the help of a straight snip of suitable dimensions.
5. Remove the sharpened edge and fold the edge of the sheet once by using a mallet (wooden hammer).
6. Repeat the same process to make more hem and join both pieces from the folded edge.

   **Result:** Single Seam Joint is made successfully.

### Safety Precautions

1. After the completion of work, we have to keep the tools in the proper place.
2. Care must be taken at the time of cutting of sheet.
3. Sharp tools must be used to avoid any injury.
4. A supporting tool must be taken while hammering.

**Figure 157.** Single Seam Joint

## Experiment 4

**Aim:** To make a Locked Seam Joint.

**Materials Required:** G.I. sheets of suitable dimensions.

### Tools Required

- Steel rule.
- Try square.
- Shear.
- Hammer.
- Mallet (wooden hammer).
- Straight snip

### Theory

**Locked Seam Joints:** It is used for the locking purpose of seam joints in sheet metalwork.

### Procedure

1. Take the measurement for cutting the sheet using a steel ruler.
2. Check the straightness of the sheet.
3. Cut the G.I. sheet strip with the help of a straight snip of suitable dimensions.
4. Now remove the sharpened edges.
5. Fold the edge of the sheet once by using a mallet (wooden hammer).
6. Make another piece the same as the above.
7. Join together from the folded edges and then the hammering should be done on the joint with the help of a mallet.

   **Result:** The seam joint is locked perfectly and the locked seam joint is made successfully.

### Precautions

1. After the completion of work, we have to keep the tools in the proper place.
2. Care must be taken at the time of cutting of sheet.
3. Sharp tools must be used to avoid any injury.
4. A supporting tool must be taken while hammering.

## Experiment 5

**Aim:** To make Wired Edge Hem Joint.

**Materials Required:** G.I. sheet of suitable dimension.

### Tools Required

- Steel rule.
- Shear.
- Try square.
- Hammer.

### Theory

- Mallet (wooden hammer).
- Straight snip.
- A wire.

**Wired Edge Joints:** Sheet metal edges are enabled by this joint to join the pieces along with them.

### Procedure

1. Take the measurement for cutting the sheet using a steel ruler.
2. Check the straightness of the sheet.
3. Cut the G.I. sheet strip with the help of a straight snip of suitable dimensions.
4. Now remove the sharpened edges.
5. Fold the edge of the sheet in a circular shape by using a mallet.
6. Put the wire in the circular region and beat it again with a mallet.

   **Result:** Making of the wired edge hem is completed.

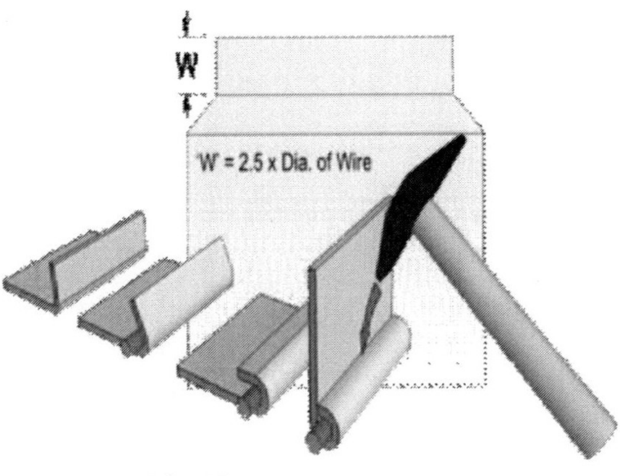


**Figure 158.** Edge Hem

### Safety Precautions

1. The thumb should be kept farther than a hammer.
2. Care must be taken at the time of cutting of sheet.
3. Sharp tools must be used to avoid any injury.
4. A supporting tool must be taken while hammering.
5. Safety must be taken while using wire and snip.

## Sheet Metal Shop Objective Questions

1. Bent snip is used for the :
   a) Removing the Burr
   b) Making Holes on the Sheet
   c) Bending the Sheet
   **d) Cutting the Sheet along Curved Lines**

2. Copper and Brass sheets are :
   a) Hard and Brittle
   **b) Malleable and Ductile**
   c) Malleable and Brittle
   d) Hard and Ductile

3. Which of the following is the purpose for using a hand groover in sheet metal working?
   a) **To Lock Grooved Seam**
   b) To Unlock the Grooved Seam
   c) To Strike the Sheet
   d) To Hold the Sheet in Position

4. Which tool is used for flattening the metal around the punched hole.
   **a) Ball Pane Hammer**
   b) Riveting Hammer
   c) Setting Hammer
   d) Sledge Hammer

5. Which of the following type of joint is used in which the end of a sheet is placed over the end of another sheet and joined together?
   a) **Lap Joint**
   b) Butt Joint
   c) Knocked Up and Joined
   d) Grooved Seam Joint

6. Sheet metal work is carried out only on metal sheets which are:
   a) Cast
   b) Forged
   **c) Rolled**
   d) None of the Above

7. To cut the sheet metal at 45 or 90 degrees, which machine is used?
   a) Drilling
   **b) Notching**
   c) Bending
   d) Welding

8. To make a small indentation (central hole) in sheet metal_______ is used.
   a) Pencil
   **b) Central Punch**
   c) Needle
   d) Scribbler

9. To fasten the sheet metal, a steel nail is used for _______.
   a) **Rivet Pliers**
   b) Pliers
   c) Punch
   d) Stapler

10. Sheet is made by the combination of carbon and _____.

    a) **Iron**
    b) Wood
    c) Aluminum
    d) Rubber

## Sheet Metal Shop Subjective Questions

1. Write the types and applications of punches and shears?
2. Describe the common sheet metal operations and processes with a suitable sketch.
3. Give a detailed list of hand tools used in sheet metal work.
4. Name the metals commonly used in sheet metal work.

# 4

# Welding

## Introduction

Welding is a fabrication process in which two or more materials or metal pieces are joined or fused with the application of heat, pressure, or both allowing them to cool and this causes fusion. Besides the work piece, a filler material is also sometimes added to the joint to form a pool of molten material that cools to form a joint based on welding and that will be stronger than the parent metal.

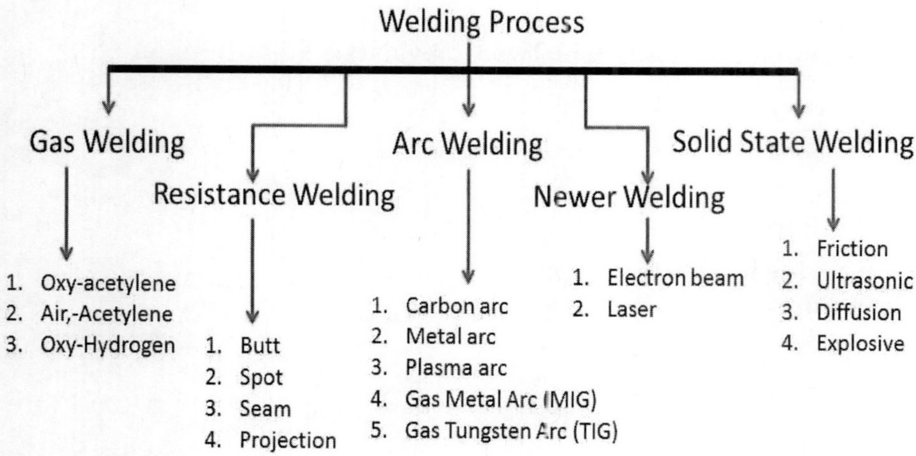


**Figure 159.** Welding Process

## Arc Welding and its Types

It is a process of joining two metals using an electric current generated by the power supply, which gives the required amount of heat to metal pieces through which the metals melt and when the metal cools, which results in the binding of the two metals. In the case of this type of welding, the power supply is given, which creates an electric arc between the electrode and the metal rod. Both A.C. or D.C. power supplies can be used in arc welding and both consumable and non-consumable rods can be used.

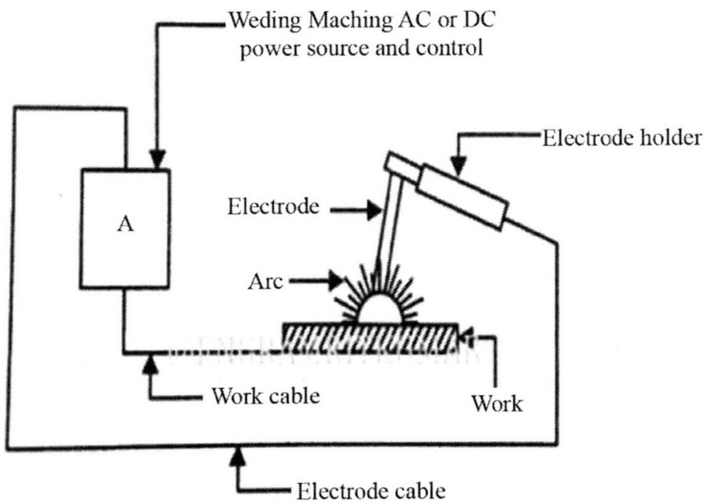


**Figure 160.** Basis Arc Welding Circuit

## Arc Welding is Classified in Two Categories:

1. Consumable arc welding processes
2. Non-consumable arc welding processes

### 1. Consumable Arc Welding Process

1. **Shielded Metal Arc Welding:** It is a kind of arc welding. The arc itself is shielded with a few flux materials. It is covered in this sort of via in order that the metallic must no longer come out of the welding. It is one of the oldest and quickest welding procedures. The strong weld may be shaped and protection may be accomplished at the right charge.

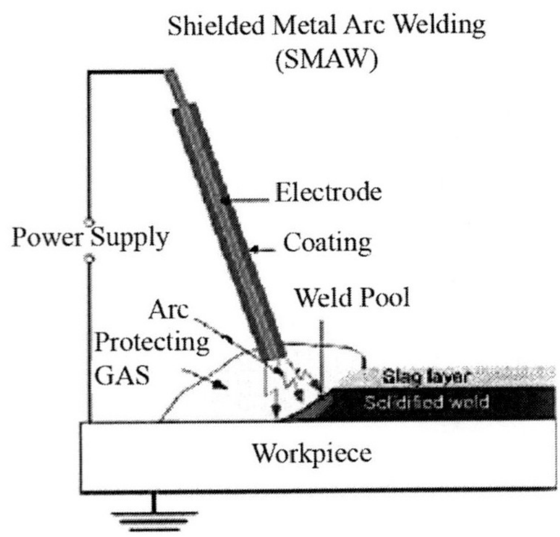


**Figure 161.** Shielded Metal Arc Welding

2. **Gas Metal Arc Welding:** It is the process in which the materials are joined together with the means of gas flame. It is the oldest method of welding and is widely used in melting metals. In this process, heat is generated by the combustion of combustible gases with oxygen. The common gases used for gas welding are acetylene, commercial LPG, hydrogen, etc. In this process, a filler rod is used, which due to the high temperature generated by the burning of gases and the filler metal melts and feeds into the gap between the joint, forming a firm weld joint. The temperature of the flame can easily be controlled by changing the proportion of oxygen and the other gas used.in this process, the combustion takes place at the outlet of the torch or the nozzle.

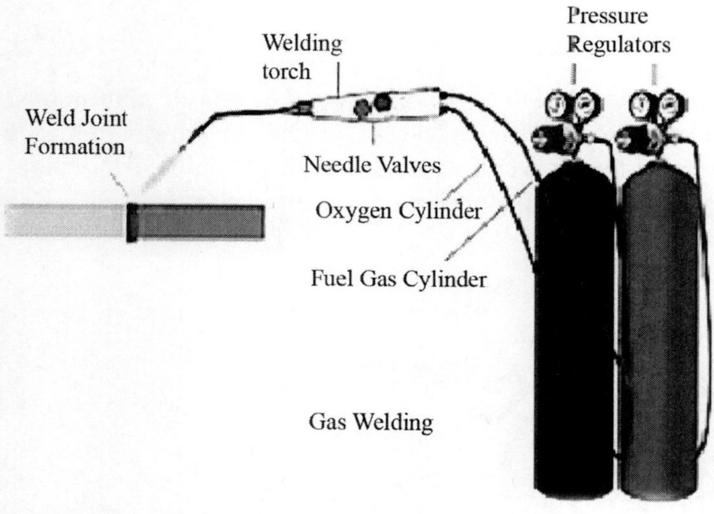


**Figure 162.** Gas Metal Arc Welding

## Advantages

- It has a higher arc time.
- Higher charge of deposition.
- Easily estimates the hassle of slag removal.
- It may be automated without problems.

**3. Flux-Cored Arc Welding:**

It is of two types:

i) Self- shielded flux- cored arc welding

ii) Gas-shielded flux-cored arc welding

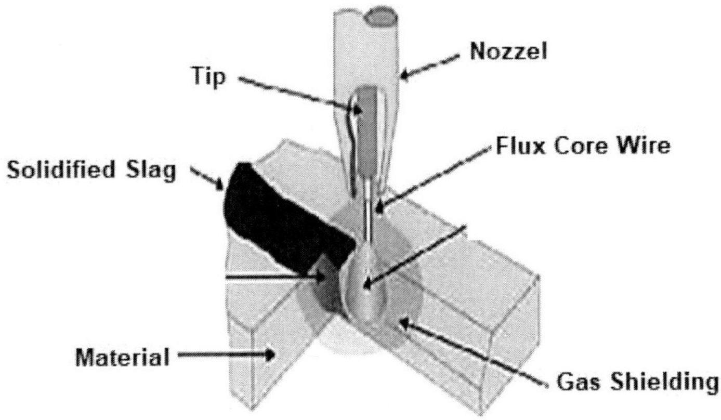


**Figure 163.** Flux-Cored Arc Welding

4. **Submerged Arc Welding:** It additionally makes use of continuous, consumable naked twine electrodes with arc protection via way of means of a cowl of granular flux. It is utilized in the metal fabrication of structural shapes and seams for big-diameter pipes, tanks and stress.

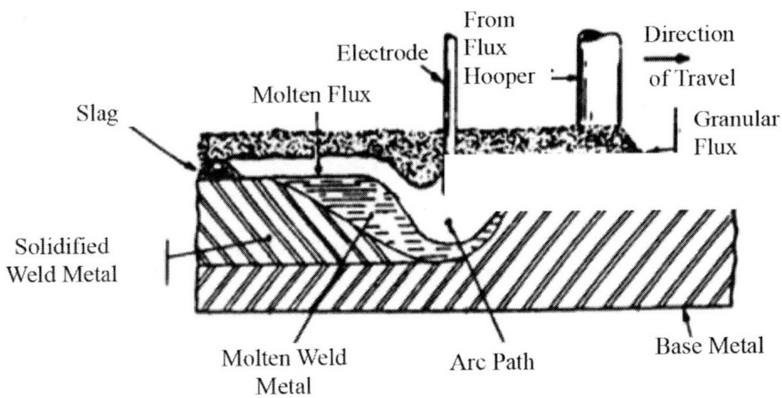


**Figure 164.** Submerged Arc Welding

## 2. Non-Consumable Arc Welding

1. **Gas Tungsten Arc Welding:** It additionally makes use of a non-consumable tungsten electrode and an inert. It additionally makes use of continuous, consumable naked twine electrodes with arc protection via way of means of a cowl of granular flux. It is utilized in the metal fabrication of structural shapes, seams for big-diameter pipes, tanks and stress, fuel line for arc welding. Melting factor of tungsten is 3410 $^0$C (6170 F). Its is used in aluminum and stainless metal.

## Advantages

- No, submit welding cleansing due to no flux.
- High pleasant weld.

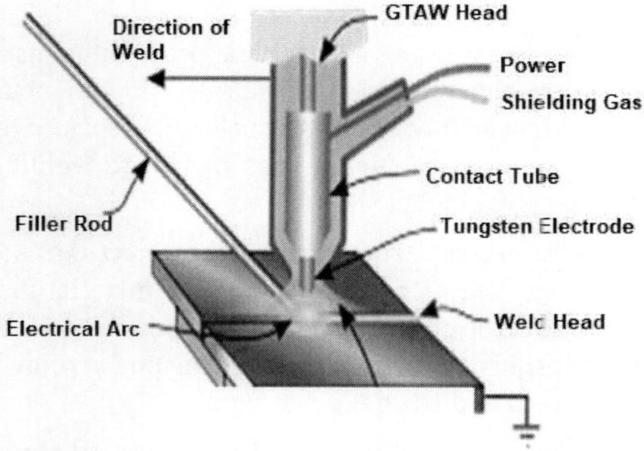


**Figure 165.** Gas Tungsten Arc Welding

## Disadvantages

- Slow and expensive in comparison to consumable arc welding.

2. **Plasma Arc Welding:** It is a unique shape of fuel line tungsten arc welding wherein a restrained plasma arc is directed at the weld area.

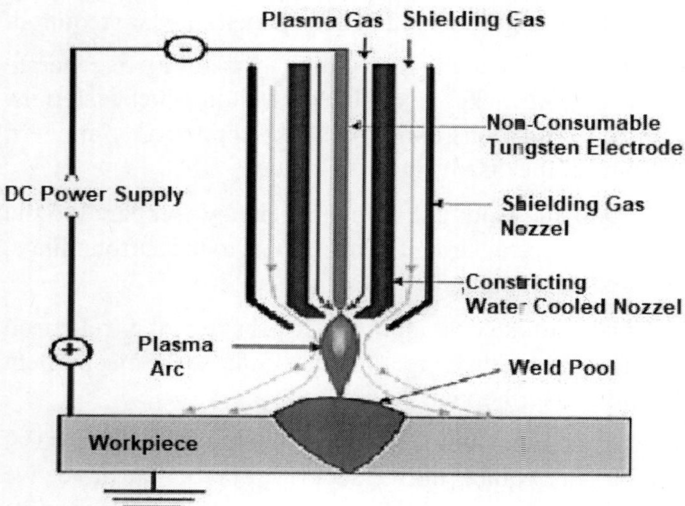


**Figure 166.** Plasma Arc Welding

## Advantages

- It offers an excellent arc balance and fantastic weld pleasant.
- It has excessive tour speeds.
- It has higher penetration control.

3. **Carbon Arc Welding**: Carbon arc welding is a welding process that uses an electric arc between a carbon electrode and the workpiece to join metal components. It is an older welding method that has been largely replaced by more advanced and efficient processes such as gas metal arc welding (GMAW) and shielded metal arc welding (SMAW).

   In carbon arc welding, the arc is generated between a carbon electrode and the workpiece. The carbon electrode is made of a graphite material that can withstand high temperatures. The electrode is held in a welding torch or holder and is connected to the positive terminal of the welding power source, while the workpiece is connected to the negative terminal.

   During the welding process, the electrode is brought close to the workpiece, and an electric current is passed through the circuit. The intense heat generated by the electric arc melts the metal at the welding point, creating a pool of molten metal. The operator then moves the electrode along the joint, depositing molten metal to form a weld bead.

   Carbon arc welding can be used for both ferrous and non-ferrous metals, but it is most commonly used for welding mild steel, cast iron, and certain types of non-ferrous metals. It is typically employed for heavy-duty applications where high deposition rates and deep penetration are required.

   One of the main advantages of carbon arc welding is its ability to generate very high temperatures, which makes it suitable for welding thick materials. However, it has several limitations and drawbacks compared to modern welding methods. Some of the disadvantages include:

i) **Low Efficiency:** Carbon arc welding has low efficiency compared to the GMAW or SMAW. A significant amount of heat is lost to the surroundings, leading to higher energy consumption.

ii) **Lack of Control:** Control of the heat input and the shape of the weld bead in carbon arc welding is very challenging. This can be due to the inconsistent weld quality and a higher likelihood of defects.

iii) **Safety Concerns:** Carbon arc welding produces intense ultraviolet (UV) radiation, it is harmful. This requires the welder to use appropriate protective gadgets, such as welding helmets with UV filters, to prevent eye damage and skin burns.

iv) **Electrode Consumption:** Carbon electrodes in carbon arc welding have a limited lifespan and need frequent replacement.

v) Due to these limitations and advancements in welding technology, carbon arc welding is now less commonly used in industrial applications. Its applications in specific scenarios or for specialized welding requirements.

4. **Electro-Gas Welding**: Electrogas welding (EGW) is a welding process that is used for the vertical position welding of thick metal plates. It is a variation of the gas metal arc welding (GMAW) process, specifically designed for vertical welding applications.

   In electro-gas welding, a consumable electrode wire is fed continuously into the joint being welded. The weld joint is typically prepared with a narrow groove, and the electrode wire is positioned at the bottom of the groove. The electrode wire is usually larger in diameter compared to GMAW, and it is often made of a continuous coil rather than being spooled.

   The welding operation is carried out with a welding torch that directs a shielding gas, typically a mixture of argon and carbon dioxide, around the electrode and the arc. The shielding gas protects the molten weld pool from atmospheric contamination, ensuring a high-quality weld.

   A unique characteristic of electro-gas welding is the use of a flux-cored electrode. The electrode wire contains a flux that melts and forms a protective slag over the weld pool as it is consumed. The slag helps to shield the weld from the surrounding air and facilitates the vertical welding process by supporting the molten metal.

   The electro-gas welding process is often automated, utilizing welding manipulators or mechanized systems to control the welding parameters and ensure consistent weld quality. It is commonly employed in applications where thick steel plates need to be joined together, such as shipbuilding, the construction of storage tanks, and heavy structural fabrication.

Some advantages of electro-gas welding are given below:

i) **High Productivity:** Electrogas welding allows for high deposition rates due to the larger electrode wire and continuous feeding. This leads to increased productivity compared to other welding processes.

ii) **Vertical Welding Capability:** The process is specifically designed for vertical welding, making it suitable for applications where vertical welds are required, such as in shipyard constructions.

iii) **Reduced Distortion:** The vertical welding position minimizes distortion in the welded structure, as the molten metal is pulled down under the influence of gravity.

iv) **Good Weld Quality:** Electrogas welding produces high-quality welds with good penetration and sound fusion between the base metals, thanks to the shielding gas and flux-cored electrode.

However, electro-gas welding also has some limitations and considerations:

i) **Limited Position:** Electrogas welding is primarily used for vertical welding. It may not be suitable for other positions, such as horizontal or overhead welding.

ii) **Equipment and Set-Up:** Electrogas welding requires specialized equipment, including a welding manipulator or mechanized system, to control the welding parameters. Initial set-up and equipment costs can be significant.

iii) **Flux Considerations:** The use of flux-cored electrodes introduces the need for proper handling and disposal of the welding flux. The slag produced during welding must be removed after each weld pass.

Electrogas welding is a specialized process that offers advantages for specific applications. It is important to consult with experienced welders and follow proper welding procedures to achieve optimal results when using electro-gas welding.

5. **Stud Arc Welding** : Stud arc welding, also known as stud welding, is a welding process used to attach metal studs or fasteners to a base material. It is a highly efficient and versatile method commonly used in construction, shipbuilding, automotive manufacturing, and other industries where fastening metal components is required.

   In stud arc welding, a metal stud, typically made of steel, stainless steel, or aluminum, is inserted into a specialized welding gun or stud welding gun. The stud is positioned against the base material, and an electric arc is initiated between the tip of the stud and the surface of the base material.

   The electric arc creates intense heat, melting the tip of the stud and a small area of the base material. Once the desired temperature is reached, the arc is extinguished, and the stud is pressed firmly against the molten pool on the base material. Pressure is applied to ensure a solid bond, and the metal cools and solidifies, forming a secure weld joint.

6. **Capacitor Discharge (CD) Stud Welding:** There are two main methods of stud arc welding:

   In this method, a high-voltage, low-energy electrical discharge is used to rapidly heat and weld the stud to the base material. CD stud welding is typically used for smaller stud sizes and thin materials.

   **Drawn Arc (DA) Stud Welding:** This method utilizes a controlled arc created by a welding power source to heat the stud and base material. The arc is initiated by lifting the stud slightly from the surface and striking the arc. DA stud welding is suitable for larger stud sizes and thicker materials.

Stud arc welding offers several advantages:

i) **Speed and Efficiency:** Stud welding is a fast process that can attach a stud to the base material in milliseconds. It allows for rapid production rates and reduces assembly time compared to traditional fastening methods.

ii) **Strong and Reliable Joints:** Stud arc welding creates a strong and durable weld joint that typically exceeds the strength of the base material. The welded studs provide reliable connections that can withstand high loads and vibrations.

iii) **Versatility:** Stud arc welding can be performed on a variety of base materials, including steel, stainless steel, aluminum, and other metals. It can also be used to weld studs to different shapes and sizes of workpieces.

iv) **Minimal Surface Preparation:** Stud welding does not require extensive surface preparation like drilling or tapping. As long as the base material is clean and free from contaminants, the welding process can be initiated.

However, there are a few considerations with stud arc welding:

i) **Accessible Base Material:** Stud welding requires access to the backside of the base material for the stud to be properly welded. In certain situations, such as confined spaces or areas with limited accessibility, stud welding may not be feasible.

ii) **Material Compatibility:** It is important to ensure that the stud material and the base material are compatible in terms of their weldability and mechanical properties. The welding parameters must be appropriately set to achieve a reliable weld joint.

Stud arc welding is a versatile and efficient method for fastening metal studs to base materials. It offers speed, strength, and reliability in a wide range of applications, making it a popular choice in various industries.

7. **Resistance Welding :** Resistance welding is a group of welding processes that use the application of heat generated by the resistance to electric current to join metal parts together. In resistance welding, the metal parts to be joined are held together under pressure, and an electric current is passed through the joint, generating heat at the contact points. The heat causes the metal to melt and form a weld.

Here are some common types of resistance welding:

i) **Spot Welding:** Spot welding is the most widely used resistance welding process. It involves the creation of welds at localized points, or spots, between two overlapping metal sheets or parts. Spot welding is commonly used in the automotive industry for joining sheet metal components.

ii) **Seam Welding:** Seam welding is similar to spot welding but creates a continuous weld along a seam or joint. It involves the movement of electrodes along the joint, producing a series of overlapping spot welds. Seam welding is often used in the production of pipes, tubes, and fuel tanks.

iii) **Projection Welding:** Projection welding is used when specific features or projections are present on the metal parts to be joined. The projections concentrate the heat and pressure, allowing for efficient welding. Projection welding is commonly used for joining fasteners, such as nuts, bolts, and studs, to metal sheets.

iv) **Flash Butt Welding:** Flash butt welding is used for joining two metal parts end to end. The ends of the parts are brought into contact, and an electric current is passed through them. As the current flows, the heat generated causes the ends to melt and then forge together under pressure. Flash butt welding is often used in the production of wires, rods, and rails.

v) **Upset Butt Welding:** Upset butt welding, also known as resistance butt welding, is used to join two metal parts end to end, similar to flash butt welding. However, in upset butt welding, the parts are brought into contact and heated by resistance without the use of a flashing or forging process. This type of welding is commonly used for smaller-diameter wires and rods.

Each type of resistance welding has its specific advantages and applications. They all offer benefits such as speed, efficiency, and localized heat generation, making them suitable for various industrial welding needs. The selection of the appropriate resistance welding process depends on factors such as the materials being joined, the required joint strength, and the desired production rate.

### Types of Flames

1. **Neutral Flame**: The neutral flame is the most commonly used welding flame. It consists of an inner cone and an outer envelope. The inner cone has a bluish color and is surrounded by a lighter, slightly luminous outer envelope. The neutral flame indicates a well-balanced mixture of fuel gas (such as acetylene, propane, or natural gas) and oxygen. It provides a stable heat source with optimal temperature and gas coverage for most welding applications.

2. **Carburizing Flame**: The carburizing flame has a longer inner cone compared to the neutral flame. It appears slightly yellowish or reddish and is surrounded by a bluish outer envelope. This flame is rich in fuel gas and

has a limited amount of oxygen. It is used for processes like flame hardening or when a localized increase in carbon content is desired, such as in case hardening.

3. **Oxidizing Flame**: The oxidizing flame has a short inner cone and a pronounced blue outer envelope. It is rich in oxygen and has a limited amount of fuel gas. The oxidizing flame produces a high heat output and is used for applications like cutting, gouging, or welding certain non-ferrous metals. It provides good control over the oxidization of the base metal.

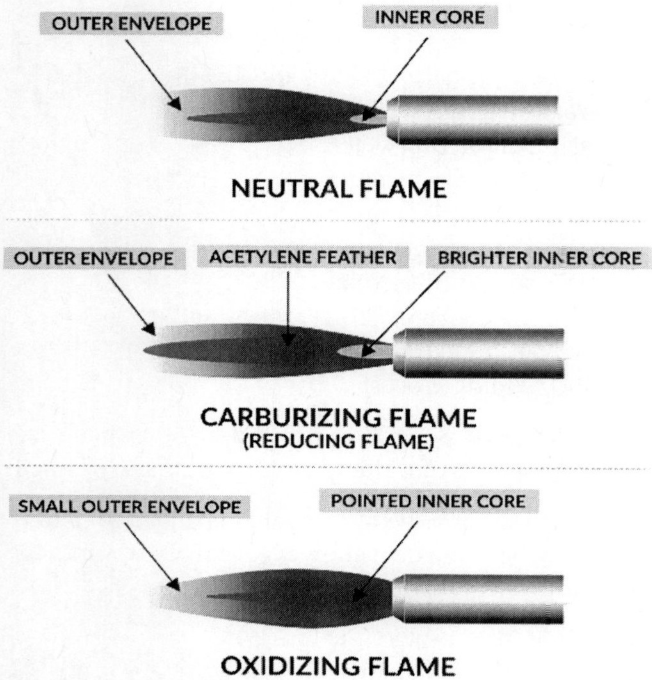


**Figure 167.** Types of Welding Flames

## Tools Used in Welding

- A Welding Table or Workbench.
- The workpiece (Parent metal). Material
- Safely goggles.
- Welding Helmet.
- Hand Gloves.
- Aprons.

- A Chipping Hammer.
- Wire Brush.
- Clamp (mostly C-Clamps are used).
- An Angle Grinder.
- Tape measures.
- Power supply (A.C. Or D.C).
- Welding electrode.
- Electrode holder.

## Uses of Tools Required

1. **A Working Table or A Workbench:** A welding table is used to provide a stable surface on which the welding process can be done easily. Another benefit of a welding table is getting a comfortable place. Its size can range from 20"x40" to 6.5"x13". It is made up of thermal resistance material, which isolates the surrounding place through light emitted and heat that is generated during the welding process.

**Figure 168.** Working Table

2. **Work Pieces (Parent Metals):** It is a piece of metal or other materials which are being joined with the help of welding or the parts that are joined.
3. **Filler Materials:** It is a material or a metal piece that is added to help to form a joint. It is heated, liquefied, and then allowed to melt to flow into the space between the parent metals.
4. **Safely Goggles:** They are used to protect the eyes from the intense rays (Ultraviolet rays, Infrared rays, etc.) that are generated during welding operations; otherwise, it may cause serious problems in the eyes, like it can burn the cornea etc.

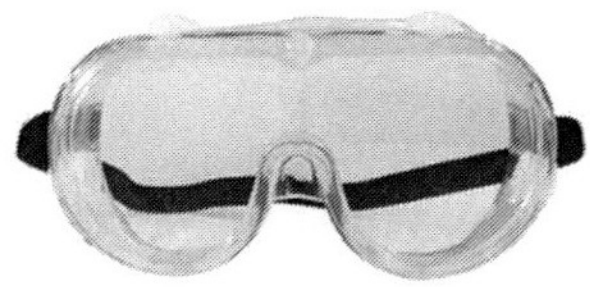

**Figure 169.** Welding Helmet

5. **Welding Helmet:** They are a type of face cover that covers our whole face, also the neck, from the sparks and the ultraviolet rays or infrared rays that are generated during the welding process. It is mainly used in arc welding (such as shielded metal welding, gas metal arc welding and tungsten welding.)

**Figure 170.** Safely Goggles

6. **Hand Gloves:** The welding or hand gloves are used to protect the hands from arc generated, hot parts of the metal, electric shocks, flames and the metal's sharp edges.

7. **Aprons:** Aprons are worn by workers to protect or safeguard them from the harmful flames, arcs, sparks, and chemical splashes generated during welding.

8. **A Chipping Hammer:** It is a tool used to remove slag from the weld during the welding process. The wide part of the hammer is used to remove the slag in the direction of the job. It is mostly used in arc welding processes. It is mostly made up of stainless steel.

**Figure 171.** Hand Gloves

**Figure 172.** Aprons

**Figure 173.** Chipping

9. **Wire Brush:** It is used to remove dust, slag, and rust or simply to clean the welding surfaces (for creating a better conductive area for electrical conduction). The bristles of the wire the brush is made up of stainless steel.

**Figure 174.** Wire brush

10. **Clamps:** It is used to hold the two metals or materials tightly so that they can be welded together easily. They are most necessary in large metals or metal sheets that are difficult to hold with our hands. In welding processes, the C-clamp (also called G- clamp) is generally used. It can be used when there is a requirement of a certain pressure to hold workpieces together. Its disadvantage is that it takes a lot of time to release the metal piece and readjust.

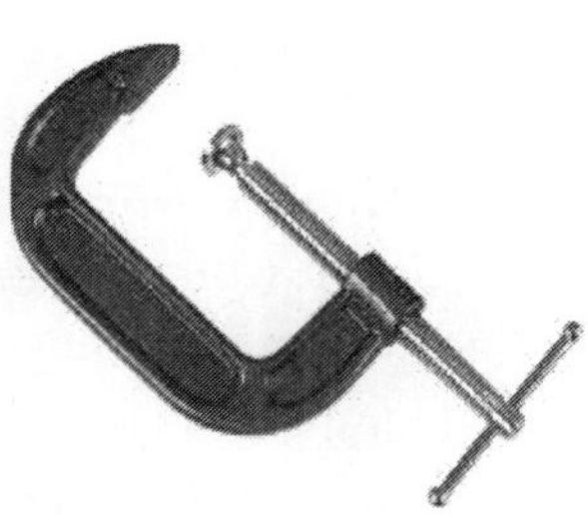

**Figure 175.** Clamps

**Figure 176.** Angle Grinder

11. **Angle Grinder:** An angle grinder is a handheld tool that consists of a consumable stone disc or blade that rotates or spins at high speeds and is used in cutting metal, cleaning, or finishing welds

also, for polishing and grinding. It can also be used for smoothing welds and also for removing rust.

12. **Tape Measures:** The tapes are used by the welders to measure the length of the metals and the curves etc. The best tape measures for welder's use are made up of plastic, fiberglass, metal strips, or cotton ribbon.

13. **Power Supply (A.C. or D.C.):** The power source is used to provide a sufficient amount of heat which can help the metal pieces to produce a stable arc. It is used for relatively welding to complicated welding processes. The transformer is used, which changes high voltage, low ampere power to low voltage, high amperage. The A.C. supply (usually used in all arc welding processes is obtained from a transformer, while for the D.C. supply, the power is generated from generators driven by an electric motor.

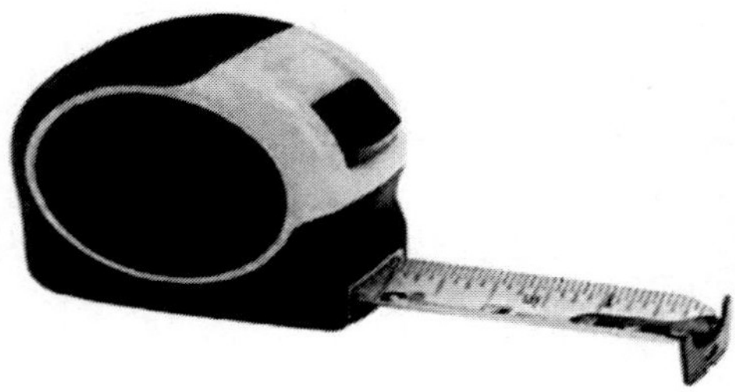

**Figure 177.** Measuring Tape

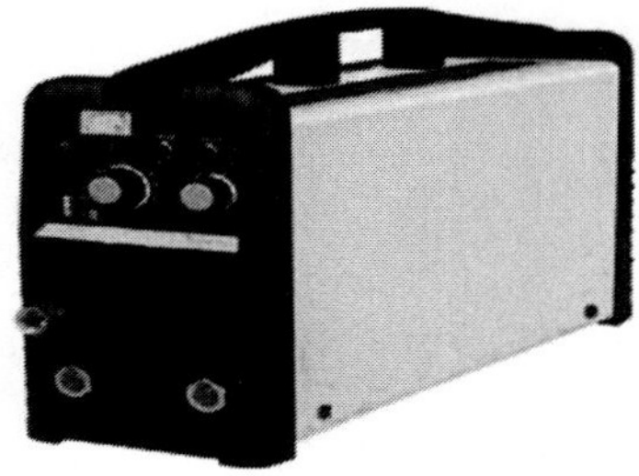

**Figure 178.** Transformer

14. **Welding Electrode:** It is an electrode with or without the coating of metal or alloys. The electrodes are the following types :
    i) **Consumable Electrode:** This is a type of electrode that melts and acts as filler metal for welding arc.
    ii) **Non-Consumable Electrode:** It is a type of electrode that does not melt while an additional filler metal is required to be added to form a welding arc.
    iii) **Bare Electrode**: This is a type of electrode of metal or alloy without any flux coating over it.
    iv) **Flux-Coated Electrode**: This is a type of electrode with flux coating covered over the metal or alloy. The cover or Flux acts as a shield to the welding arc.

15. **Electrode Holder:** The electrode holder is used to hold the electrode and to conduct current to it. It is connected at the end of the welding cables. The electrode is said to be the best electrode when it is light in weight, should not become hot during welding operations, and is easy to handle.

The jaws of the electrode holder are made up of insulators or are insulated to avoid electric shocks. Their size varies from 150 to 500 amps.

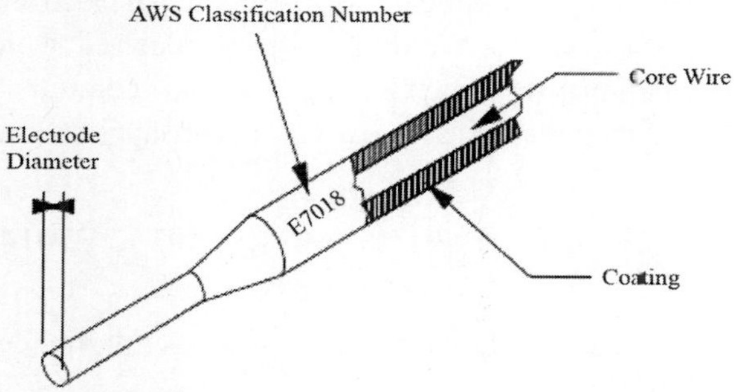


**Figure 179.** Welding Electrode

**Figure 180.** Electrode Holder

## Precautions

1. Always wear helmets or shields which protect you from UV radiation, hot slag, chemical burns, etc., which can harm your body.
2. Wear fire-resistant clothes which protect you from heat and fire radiation that is generated during welding processes. Do not use synthetic clothe used leather or cotton clothing.
3. Wear insulated gloves and boots which protect you from heat and also protect you from electric shocks. Try to wear rubber-soled shoes or flame-resistant shoes.
4. Before starting welding, it is important to get training from someone who is good at it; otherwise, E-Courses provide an east training opportunity.
5. Avoid welding in a place that is not free from flammable materials as the flammable may catch fire from the flames or head generated during welding and which can affect you.
6. Fire extinguishers must be ready for use whenever it is needed.

## Types of Welding Electrode

There are various types of welding electrodes, also known as welding rods, available for different welding processes and applications. The choice of electrode depends on factors such as the base metal, welding position, desired weld characteristics, and the specific welding process being used. The some commonly used welding electrodes, along with their materials and recommended applications are given below:

1. **Mild Steel Electrodes (types AWS E6010, E6011, E6013, E7014, E7018):**

   **Materials:** Mild steel electrodes are typically made of low-carbon steel with a coating that contains flux.

   **Applications:** These electrodes are versatile and widely used for welding mild steel structures, general fabrication, maintenance, and repair work.

2. **Stainless Steel Electrodes (types AWS E308L, E309L, E316L):**

   **Materials:** Stainless steel electrodes are made of alloys containing chromium and nickel, which provide corrosion resistance and heat resistance.

   **Applications:** They are used for welding stainless steel, such as in food processing equipment, chemical plants, and marine applications.

3. **Cast Iron Electrodes (types AWS E7018-1):**

   **Materials:** Cast iron electrodes have a high iron content with added elements such as nickel and ferrosilicon.

   **Applications:** They are specifically designed for welding cast iron components, including engine blocks, pump housings, and machinery parts.

4. **Low Alloy Steel Electrodes (types AWS E8018, E9018):**

   **Materials:** Low alloy steel electrodes contain additional alloying elements such as manganese, nickel, and molybdenum for enhanced strength and toughness.

   **Applications:** These electrodes are used for welding low alloy steels, such as in structural fabrication, pressure vessels, and high-strength applications.

5. **Aluminum Electrodes (types AWS E4043, E5356):**

   **Materials:** Aluminum electrodes are made of aluminum alloys with specific compositions for good weldability.

   **Applications:** They are used for welding aluminum components in industries like automotive, aerospace, and construction.

6. **Tungsten Inert Gas (TIG) Electrodes ( types AWS and EWP):**

   **Materials:** TIG electrodes are typically made of pure tungsten or tungsten alloys.

   **Applications:** TIG electrodes are used in TIG welding to produce high-quality welds on various metals, including stainless steel, aluminum, and copper alloys.

7. **Flux-Cored Electrodes (types AWS E71T-1, E71T-11):**

   **Materials:** Flux-cored electrodes consist of a metal core filled with flux, eliminating the need for external shielding gas in some cases.

   **Applications:** They are commonly used in construction, shipbuilding, and other industries where high deposition rates and deep penetration have occurred.

## Experiment 1

**Aim:** To join the two metals or work pieces by arc welding as a Lap Joint.

**Materials Required:** Metal pieces.

### Tools Required

- Electrode.
- Electrode holder.
- Chipping hammer.
- Power Supply (A.C. or D.C.).
- Wire brush.
- Angle grinder.
- Gloves and Apron.
- Clamps.
- Welding helmet and goggles.
- Filler metal (If required).

### Theory

**Welded Lap Joint**: It is a type of joint in which one part of metal or its end or edge overlaps some part of the other metal and they are joined together through welding. These are mainly used when two metals of different thicknesses are to be joined. These joints are easy to prepare but they are less than the base metal.

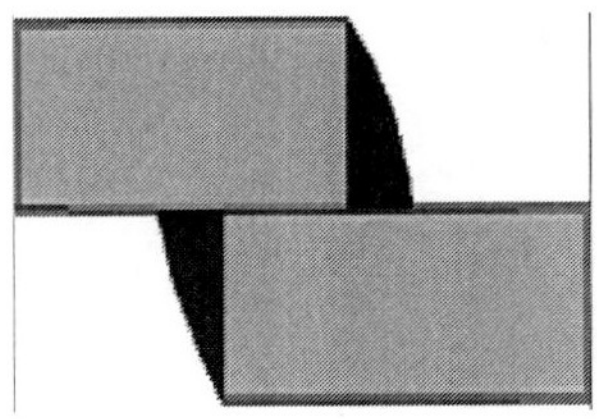

**Figure 181.** Lap Joint

### Procedure

1. First of all, take two workpieces of a certain dimension and clean them thoroughly so that no dust, rust, or anything remains on the surface of the metal piece.
2. Using the angle grinder, remove the sharp edge of the work piece and also grind it to prepare the work piece for welding.
3. Now place the two work pieces on the work table such that they form a lap joint with the required portion above as given.
4. Now hold the electrode with the help of the electrode holder and set a certain value of power generated from the power supply.

5. Fasten the clamp to the welding table.
6. Wear a shield, gloves and apron for protection from the harmful rays generated during welding.
7. The electric arc produced will melt the metal pieces and then join them together through arc welding.
8. After wearing the gloves hold the overlapped pieces the work pieces and the arc struck are tacked at both of their ends to form a tack weld.
9. Check the alignment of the lap joint as they are in the right alignment as needed or not if it is not so then reset the alignment.
10. Also, check the tack-welded pieces and reset them if required.
11. Now carry out the welding throughout the whole length say on both ends.
12. Clean the welded surface by removing slag, spatters, etc.

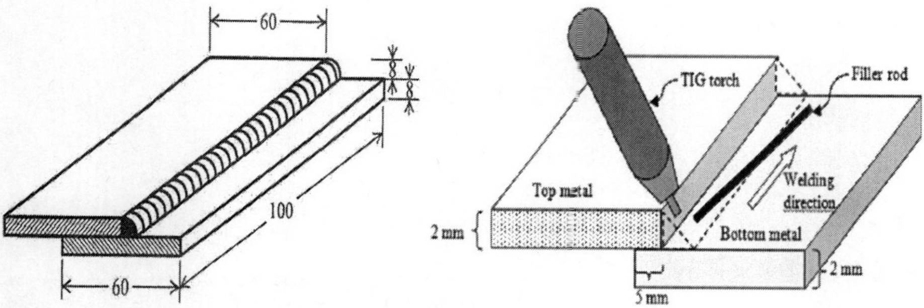


**Figure 182.** Final Result of Lap Joint

## Result

The two metal pieces are joined together by arc welding forming a lap joint.

## Precaution

1. Do not wear gloves while using the angle grinder.
2. Always wear gloves and face shields or helmets while welding, for protecting the eyes, face, and whole human body.
3. Try to maintain a constant arc length during welding.
4. Try to wear fire-resistant clothes in order to protect from electric shocks.
5. Keep fire extinguishers ready so that they can be used when required.

## Experiment 2

**Aim:** To join the two metals or work pieces by arc welding as a Butt Joint.

**Material Required:** Two metal pieces or plates.

### Tools Required

- Power supply.
- Electrode.
- Electrode holder.
- Angle grinder.
- Apron and Gloves.
- Face shield or Helmet.
- Goggles.
- Chipping Hammer.
- Wire brush.
- Clamp.
- Filler metal (If required) etc.

### Theory

**Welded Butt joint:** Butt joints are formed when two structures of metal are placed connected through their edges and then they are joined through welding; this process is called butt joint welding. It is most commonly used for the fabrication of structures. It is simple to prepare a butt joint, but the defect that comes in welded butt joints is cracking or penetration.

**Figure 183.** Butt Joint

### Procedure

1. Firstly clean the metal surface by removing the rust, grease, oil, etc.
2. Now take both metals (of the desired size) close to each and beveled them forming V-groove between them.
3. Wear the apron, gloves, helmet and other protective things in order to protect your body from the harmful radiation and flames generated during welding.
4. Now connect the whole arrangement to the power supply and on the power supply.

5. Hold the electrode by the electrode holder and then clamp the ground clamp to the surface plate and now the parent metals are placed on it.
6. The power generated from the power supply will melt the metal, which initiates the arc either by the striking method or by the touch and drag method.
7. Maintain the gap of approx 3mm between the plates and also to the electrode in order to avoid improper arc length.
8. Now remove the slag present on the metal surfaces by using a wire brush and chipping hammer.
9. Repeat the above process until the groove is fully filled by the weld.

## Result

The two metal pieces are joined together by arc welding to form a butt joint.

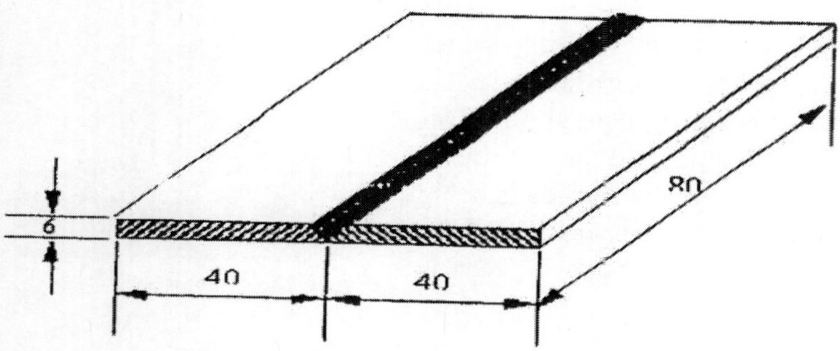


1. Single 'V' Butt Joint

**Figure 184.** 2 Plates Joined Butt Joint

## Precautions

1. Always wear gloves, shields, and an apron to avoid the effect of UV rays and other harmful rays.
2. Try to wear insulated clothing and shocks (rubber-soled) in order to avoid electric shocks.
3. Choose an area where the flames-catching materials are minimum.
4. Maintain the constant arc length.

## Experiment 3

**Aim:** To join the two pieces of metal by arc welding as a T-Joint.

**Materials Required:** Two metal pieces.

### Tools Required

- Power supply.
- Electrode holder.
- Goggles.
- Gloves.
- Wire brush.
- Filler metal (If required) etc.
- Electrode.
- Apron.
- Shields or Helmet.
- Chipping Hammer.
- Clamp.

### Theory

**Welded T-Joint:** It is a type of joint in which two metal pieces are joined together in such a way that they intersect each other at 90°. This forms the center of the plate in the shape of a T-Joint. It is mostly used when a pipe is being welded on the base metal surface. The main defect in this type of joint is that lamellar tearing may occur.

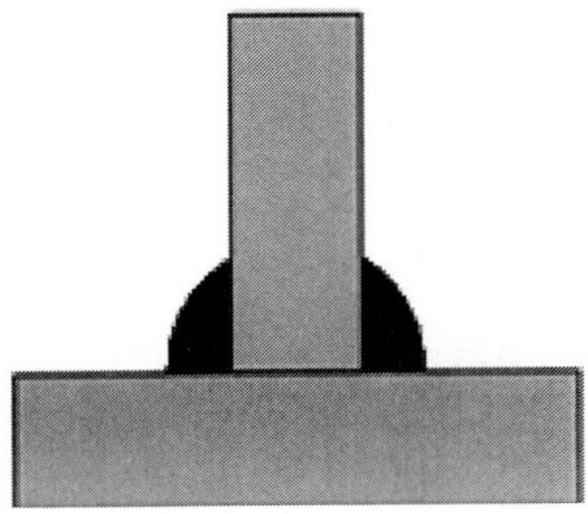

**Figure 185.** T-Joint

### Procedure

1. First, clean the welding surface in order to remove the oil, grease, rust, and dust, etc., and also grind the edges through an angle grinder if required.
2. Connect the terminal to the power supply and now hold the electrode with the help of the electrode holder. Also, clamp the ground clamp to the ground plate so that the work piece can be placed on it for welding.
3. Wear gloves, goggles, etc, to protect yourself, and then switch on the electric power supply.
4. The two metal pieces are arranged in such a manner that they form a T-Joint at their center and then plated are to be welded in the position of overlapping and tag welding is done to restrict the movement from one end.
5. The electricity generated from the power supply will produce heat which will melt the electrode joining the two metal pieces.

6. Now clean the welded surface in order to remove the slag through a wire brush and also with the use of a chipping hammer.

## Result

The two metal pieces are joined together through arc welding forming T-Joint.

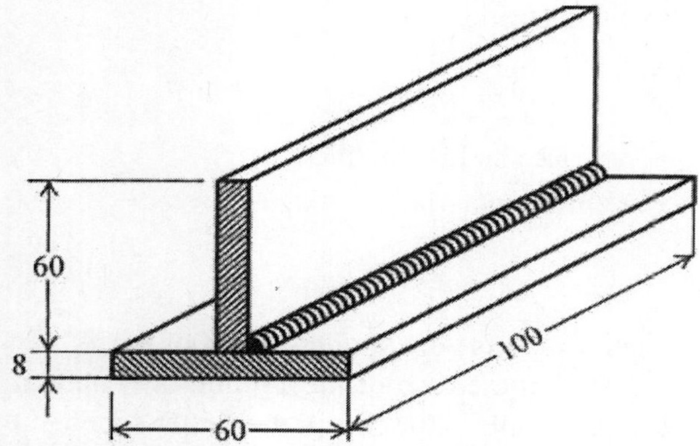


**Figure 186.** Obtained T Joint

## Precautions

1. Wear gloves, aprons, etc., to avoid the effects of flamed and harmful radiation generated during welding.
2. Try to wear insulated shoes and clothes to overcome the problem of electric shocks.
3. Maintain constant arc length.

## Experiment 4

**Aim:** To join the two pieces of metal through arc welding to form a CornerJoint.

**Materials Required:** Two metal pieces

### Tools Required

- Power supply.
- Electrode holder.
- Chipping hammer.
- Goggles.
- Wire brush.
- Electrode.
- Apron.
- Gloves.
- Face shield or helmet.
- Filler material (if required) etc.

### Theory

**Welded Corner Joint**: This type of joint is very similar to that of the T-Joint only difference between them is that in the case of the T-Joint, one metal is joined in the middle of the other metal, while in the case of the corner's joint, the metal is placed at the corner of the other metal piece forming L shape and then both the metal pieces are joined together through welding.

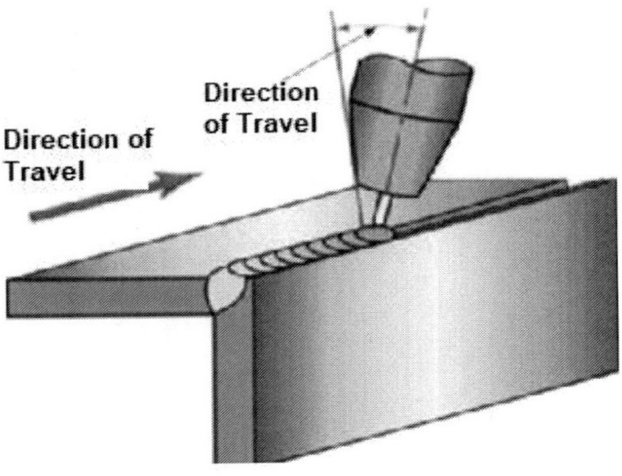


**Figure 187.** Corner Joint

### Procedure

1. Clean the metal surface in order to remove the oil, dust, grease, oil, etc.
2. Clamp the ground clamp to the surface plate and then the workpiece is placed on it for welding.
3. Hold the electrode with the help of an electrode holder. Wear gloves, goggles, etc. To avoid any harm from the flames and harmful rays.

4. Take two metal pieces close to each other in such a way that the metal piece is at the corner of the other metal piece forming an L shape and U or V-shaped grooved.
5. Now on the power supply which generates power which produces heat through which the metal piece melts and both the parent metal pieces are joined together by welding.
6. Now clean the welded surface using a wire brush and chipping hammer to remove slag.

## Result

The two metal pieces are joined together through arc welding forming Corner Joint.

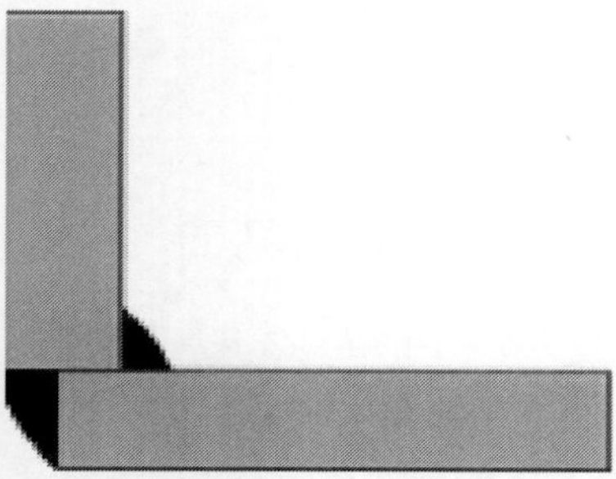

**Figure 188.** Obtained Corner Joint

## Precautions

1. Wear an apron, goggles, helmet etc. to avoid the flames and harmful radiation generated during welding which can harm our eyes, face etc.
2. Try to wear insulated clothes and rubber shoes to protect yourself from the electric shocks.
3. Maintain constant arc length.
4. Fire extinguishers must be ready so that they can be used when required.
5. Do not do the welding process where fire-catching objects are present.

## Experiment 5

**Aim:** To join two pieces of metal by gas welding to form a Lap Joint.

**Materials Required:** Two metal pieces.

### Tools Required

- Welding torch.
- Oxygen cylinder.
- Fuel gas cylinder.
- Pressure regulator.
- Goggles.
- Gloves.
- Filler rod.
- Chipping hammer and wire brush.

### Procedure

1. First of all clean the metal plates and remove the dust grease, oil, dirt, etc.
2. Now connect the pieces of equipment carefully. Connect the oxygen cylinder and the fuel gas cylinder with the welding torch through the pressure regulator.
3. Now supply the regulated pressure of oxygen and gas fuel cylinder to the torch where they mix properly. Flames can be calculated with the help of the valves present in the welding torch and now set the flame according to the welding condition required.
4. Take the metal pieces and hold them together forming a lap joint and then move the welding torch along the line where the joint is to be made.
5. The heat produced from the nozzle by the combustion of the gases melts the interface of the metal or the filler metal that is being used, then the melted part will fill the space between the metal pieces and the joint is made.
6. Now Clean the welded surface using a wire brush and chipping hammer.

### Result

The metal pieces are joined together by gas welding forming a lap joint.

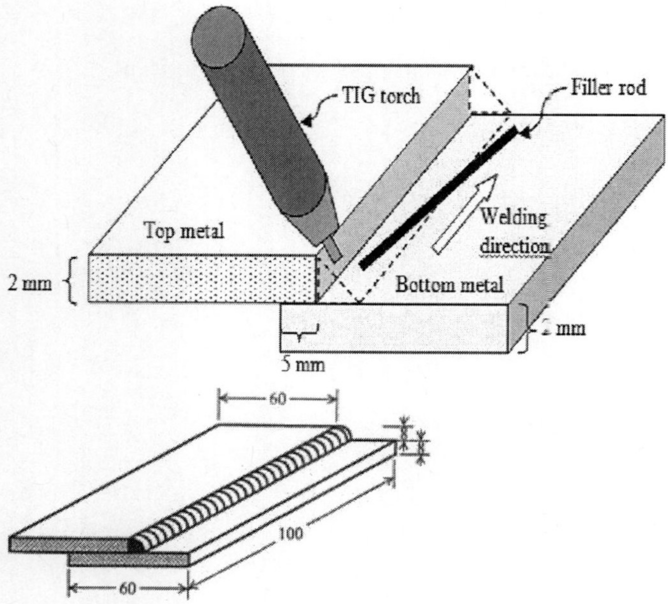


**Figure 189.** Lap Joint

## Precautions

1. The fuel gas cylinder (in the case of acetylene) is explosive so take care of that.
2. Check the leakage of the cylinders as they can leak or may harm us.
3. Protect the cylinders from the flames and sparks generated during welding.
4. After finishing welding, must check whether the cylinder valves are closed or not.
5. Try to open the cylinder valves gently and slowly to avoid the sudden increase in the pressure which may cause an explosion.

## Experiment 6

**Aim:** To join two pieces of metal by gas welding, forming a Butt joint.

**Materials Required:** Two metal pieces.

### Tools Required

- Oxygen cylinder.
- Fuel gas cylinder.
- Welding torch.
- Gloves.
- Goggles and face shield.
- Wire brush and chipping hammer.
- Pressure regulator.
- Filler metal.

### Procedure

1. Clean the metal surface gently to remove the diet grease etc., from the metal surface and connect the oxygen cylinder and another fuel cylinder to the welding torch through the pressure regulator.
2. Now regulate the pressure of oxygen and the fuel gas is supplied to the torch where they mixed properly. Set the flames according to the requirement of heat to the welding condition.
3. Take the metal pieces and join them together with their end edges and then move the welding torch along the line of the edge where the joint is to be made.
4. The nozzle produces heat which is generated due to the combustion of the gases which melts the metal edges and also the filler metal and it is filled in the space left between the metals.
5. Now clean the welded metal surface to remove the slag collected during welding through a wire brush and chipping hammer.

### Result

The two metal pieces are joined together as a butt joint through gas welding.

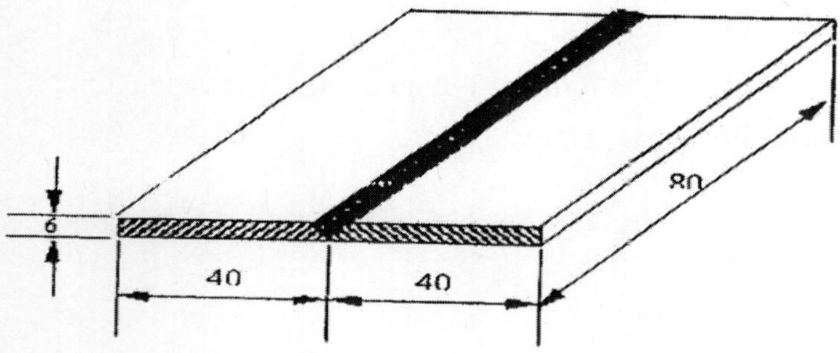


1. Single 'V' Butt Joint

**Figure 190.** Butt Joint

## Precautions

1. Always wear goggles and gloves to protect yourself from the flames that come out during welding.
2. Always check the leakage of the gases from the fuel gas cylinders.
3. After finishing welding, close the valves tightly as it can not leak any gas.
4. Fire extinguishers must be ready for use when required.

## Experiment 7

**Aim:** To join the two metal pieces forming T-Joint through gas welding.

**Materials Required:** Two metal pieces.

### Tools Required

- Oxygen gas cylinder.
- Fuel gas cylinder.
- Welding torch.
- Goggles.
- Gloves or shield.
- Pressure regulator.
- Filler metal.
- Chipping hammer or wire brush

### Procedure

1. Clean the metal surface to remove the rust, dirt, grease, etc. and connect the pieces of equipment, like connecting both the cylinders to the welding torch through the pressure regulator.
2. Now regulate the pressure of oxygen and the fuel gas supply to the welding torch where they are mixed properly. The flames generated from the welding torch can be maintained or set according to the requirement of the welding process.
3. Hold the pieces of the metal and bring them together and join them in such a way that one piece of the metal is at the center of the other and 90° to each other. And then, the welding torch is moved over them along the line where the joint is to be made.
4. The heat is generated from the nozzle, which is because of the combustion of oxygen and the other fuel gas, which melts the filler metal used and also the metal pieces. The melted material is then fed into the space between the two metal pieces and the joint is made.
5. Clean the welded surface thoroughly.

### Result

The two metal pieces are joined together as a T-Joint using gas welding.

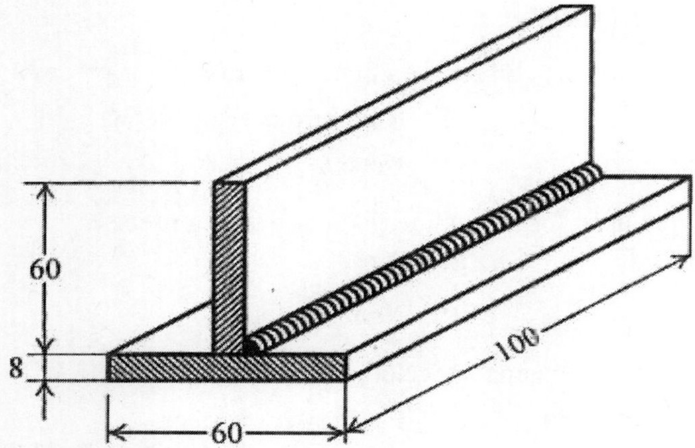


**Figure 191.** T Joint

## Precautions

1. Check the leakage from the cylinders.
2. Always keep the cylinders away from the welding area, as if they catch fire, they can harm us.
3. Fire extinguishers must be ready for use when required.
4. Try to open the valves of the cylinders slowly as opening fast may increase the pressure suddenly, and the cylinder may explode.

## Welding Shop Objectives Questions

1. Which types of metals are not good for the submerged arc welding process:
   a) Ferrous type Metals
   **b) Nonferrous type Metals**
   c) Ferri-type Metals
   d) None

2. Which type of current used for the submerged arc welding process:
   a) AC
   **b) DC**
   c) **AC and DC**
   d) **None**

3. Which of the following is consumable welding process :
   a) Gas Tungsten Arc Welding
   b) Plasma Arc Welding
   c) Carbon Arc Welding
   **d) Submerged Arc Welding**

4. In plasma arc welding the used gas is ————.
   a) Magnetized
   b) Heated
   **c) Ionized**
   d) Vaporized

5. In Argon arc welding the electrode is made of:
   a) Carbon
   b) Steel
   b) **Tungsten**
   d) No electrode is needed

6. Welding process in which coarseness is produced is called ————.
   a) Oxyfuel Process.
   **b) Resistance Process**
   c) Gasetal Arc Process
   d) Friction Process

7. When rotation motion stopped in the friction process the ————.
   a) **Temperature Increase**
   b) Temperature Decrease
   c) Pressure Decrease at Constant Volume
   d) Unchanging the Temperature

8. When rotation motion is stopped then which force works?
   a) Equatorial Force
   b) Induced Magnetic Force
   **c) Axial force**
   d) Frictional Force

9. How does the arc voltage, V depend upon the length of arc L?
   a) V = f(L)
   b) V = 1/f(L)
   c) **V = f(L2)**
   d) V = f(L)

10. The temperature obtained by the plasma arc welding process is about.

    a) 50,000^°F  b) 55,000^°F

    c) **25,000^°F**  d) 30,000^°F

## Welding Subjective Questions

1. What are the different welding techniques? Briefly explain them.
2. What are the purposes of shielding electrode coatings?
3. What are the different types of welding joints? Explain all joints with a neat sketch.
4. Differentiate between welding, soldering, and brazing processes.

# 5

# Foundary Shop

## Introduction

The procedure of making or building casting in the sand mould cavity or any other material is known as a foundry. This is the cheapest method of shaping the metal. Foundry exercise offers the technique of creating casting in moulds shaped in both sand or different material. A foundry is a manufacturing plant where castings are created by softening metal, emptying fluid metal into a shape, then, at that point, permitting it to harden. Regardless of whether we have not at any time been to a foundry or even know what one resembles, we are encircled by the metal castings. There is a decent possibility that we are perusing this under 10 feet from one.

## Moulding

Molds are negative structures that are utilized to shape projecting materials, making copies of the model (object) the form was produced using.The empty types of the form get loaded up with water (the projecting material) to create ice blocks. The 3D shapes are delivered from the form and presto! Completed cast pieces. It is formed by a pattern and is generally a cavity in a molding core. Moulding is a procedure of creating moulds by forming liquid material using a frame. There are two types of moulds based on the material used.

## Classification of Moulds

- Permanent moulds
- Temporary moulds

Permanent moulds are made of cast iron, zinc, steel, and copper alloys, whereas temporary moulds are made of sand. Moulding is principally utilized during the assembling system of plastic. Plastic is an engineered material, and to frame it into the ideal shape, diverse trim cycles are utilized. Each interaction requires the control of liquid plastic and afterward passing on it to set. Thermoplastics would then be able to be liquefied down and improved if essential, but thermoset plastics can’t be warmed.

## Casting

Casting is an essential trim interaction as it requires a minimal measure of complicated innovation. Plastic is just warmed, so it transforms into a liquid and afterward moved into a shape. It is passed on to cool and the shape is taken out. This interaction can be utilized for perplexing shapes and performed under low tension. In any case, it is a generally expected interaction utilized for making plastic sheeting, beginning from 0.5 inches thick and more prominent. In metal casting in the metallic casting procedure, metallic shapes are shaped via way of means of flowing molten metallic right into a mould cavity where in it's far cooled and later which is drawn out from the mould.

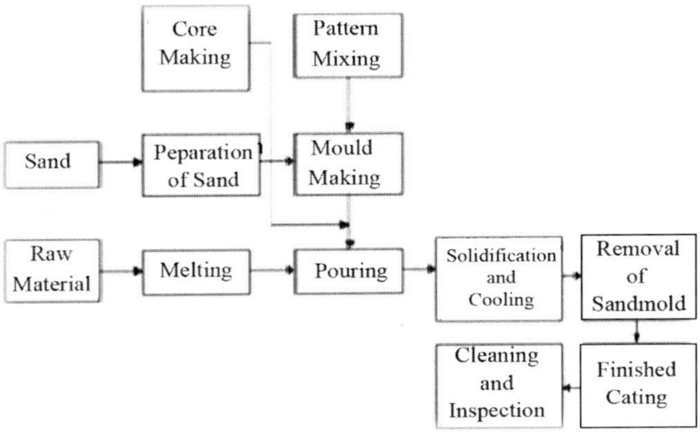


**Figure 192.** Classification of Moulds

## Steps of Metal Casting

1. **Pattern Making:** a duplicate of the component to be cast is made with the use of appropriate fabric, inclusive of wood, steel, plastic, or plaster.
2. **Mould Making:** Making of mould is a multi-step method where in style and cores are utilized to make a mould. The kind and the way the mould is made could range depending on the kind of metallic casting.
3. Metal melting and pouring: liquid is then unfrozen and poured into the hollow mould space through gravity. Then the forged is authorized to solidify earlier than the forged components are eliminated from the mould. Again,
4. The forged component elimination will range depending on the sort of metallic casting.
5. Post-processing: In this very last step, the forged steel item is eliminated from the mould, after which fettled. During the fettling, the item is wiped clean of any moulding material, and tough edges are eliminated.

## Types of Casting Process

1. **Green Sand Mould Casting** - This process involves the use of a mixture of silica sand, clay, water, and additives to create a mold. The mold is typically made of two halves and is used for casting various metals by pouring molten metal into the mold cavity.

2. **Dry Sand Mould Casting** - In this process, a mixture of sand and other additives (such as clay or resin) is used to create a mold without the presence of water. The mold is prepared using a dry sand mixture, and the casting is made by pouring molten metal into the mold cavity. Dry sand mold casting is commonly used for large and heavy castings.

3. **Shell Mould Casting** - Shell mold casting involves the use of a pre-coated sand mixture, known as shell sand, to create the mold. The mold cavity is formed by curing the coated sand mixture on a pattern or a core. Shell mold casting provides a good surface finish and dimensional accuracy and is suitable for both ferrous and non-ferrous metals.

4. **Investment Casting** - Also known as lost-wax casting, investment casting is a precision casting process. A wax pattern of the desired casting is created and coated with a ceramic material. The ceramic-coated pattern is then heated to remove the wax, leaving a hollow ceramic mold. Molten metal is poured into the mold, and once solidified, the ceramic mold is broken away to reveal the final casting.

5. **Die Casting** - Die casting is a manufacturing process in which molten metal is forced under high pressure into a mold cavity, known as a die, to produce complex-shaped parts with high dimensional accuracy and surface finish. The die-casting method is commonly used for the mass production of metal components with intricate shapes and thin walls.

6. **Centrifugal Casting** - Centrifugal casting is a casting process that involves rotating a mold or die around a central axis while pouring molten metal into it. The centrifugal force generated by the rotation helps in distributing the molten metal uniformly along the mold cavity, resulting in a dense and defect-free casting.

## Tools and Equipment

The tools and equipment needed for molding are:

Figure 193. Strike of Bar

1. **Strike of Bar:** It is also known as Strike of Edge. It is made of wood or metal. It is used to separate surplus sand and to provide a smooth and flat surface. It's far used to strike more sand from the pinnacle of the flask.
2. **Molding Board:** It is made of wood and has a smooth surface and finish. It is used to support the flask and pattern.

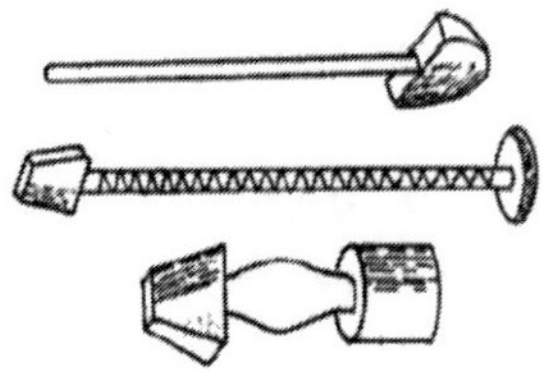

Figure 194. Rammers

3. **Rammer:** It is generally known as a hand rammer and is made of wood. It is used for Ramming the sand. The tool is used to compress the molding sand inside the flask through ramming.

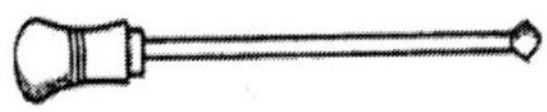

Figure 195. Vent Wire

4. **Vent Wire:** It is made of steel. Its corner has a sharp edge or sharp vent. It is used to create holes in the sand so that the gases dissolved in the hot metal may escape. It is used in making small holes with inside mildew for routing gases from the cavity.

Figure 196. Draw Spike

5. **Spike:** It is circular at one end and pointed at the other. It is used to move out the pattern from the cavity. It is used to take away the sample from the mold.

Figure 197. Trowel

6. **Trowel:** It can be made in many styles and shapes and is used to smoothen the surface. A device that is used to form and smoothen the surfaces of the mold.

Figure 198. Spruce Pin

7. **Spruce Pin:** It is made of wood. When our cavity is made it is fixed in the sand and then removed, it creates a hole so that we can pour hot metal.
8. **Riser Pin:** It is used in making a hole in the sand cope for the flowing metal to rise and is generally a tapered wooden pin.

Figure 199. Shovel

9. **Shovel:** It is used for transferring the sand into the flask. It is made of steel and provided with a wooden handle. It is also used for tempering

and mixing sand. It's far used for blending the moulding sand and used for transferring one vicinity to the other.

10. **Moulding Box or Flask:** Box that holds the moulding sand it's far open on the pinnacle and the bottom.
11. **Clamp:** clamp is used to keep a cope and drag field aligned and firm.
12. **Ladle:** the primary characteristic of the ladle is to get hold of flowing steel from the furnace and pour it into the mould.
13. **Slicktool:** They are the spoon-fashioned travels used for repairing or smoothening a mould floor and the rims after the sample has been withdrawn.
14. **Lifter:** used for repairing and completing the mould sand.
15. **Bellows:** used to remove loose or unwanted sand from the surface.

**Figure 200.** Lifter

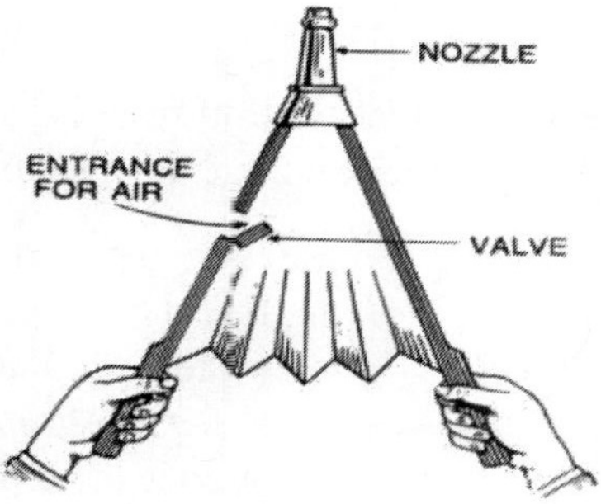


**Figure 201.** Bellows

16. **Riddle:** it's far from a display screen used to take away small portions of steel and overseas.
17. **Swab:** A swab is a brush made up of hemp fibers used to moistify the edge of sand mold that is in touch with the design surfaces before extracting the pattern(design). It can be used for covering the fluid blackening on the mold surfaces in dry sand molds and sweeping away the molding sand from the mold face in dried sand molds.
18. **Sprue Cutter:** The cutter is used in creating a hole succeeding ramming. It is the form of a hollow tube that is inserted in the sand to produce the hole.

**Figure 202.** Riddle

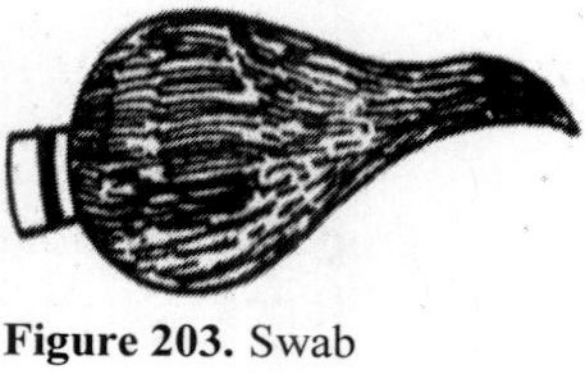

**Figure 203.** Swab

**Figure 204.** Sprue Cutter

19. **Spray-Gun:** Mold or core surfaces coated by spray-gun by spray coating of facing materials etc.

20. **Mallet:** It is the same as a wooden hammer. Mold shops, using operating the spike, bring about design then afterward wrap it for disconnection forms molding surface.

21. **Draw Spike:** It is a rod made up of steel. Used for driving into a design that is embedded in the molding sand and raps the pattern to get separated from the pattern and finally

## Materials Required in Metal Casting

- Molding sand
- Parting sand

## Molding Sand

Sand is the main element used in molding and casting. The main ingredients present in molding sand are silica, clay, sand, and additive material.

Most of the moulding is done with green sand. It contains 8% moisture. Green sand moulds are used for small and medium-sized casting. Green sand mainly contains 18- 30% clay, 6-8% water, sand and silica. Its constituents make an effective bond. Clay is essential for providing effective bonding strength to the moulding. When Moisture is added in an appropriate amount to the clay, it helps to bear the high temperature and provide.

## Types of Moulding Sand

1. **Green Sand:** It has a higher moisture content and so it can easily take any shape.

2. **Dry Sand:** It is suitable for manufacturing large castings. Finishing mould in green sand, when it is dried, is known as dry sand.

3. **Loam Sand:** This is a type of moulding sand that contains 50% of clay. In loam sand, patterns are not used.

4. **Facing Sand:** It is used to mould the face which is formed by the facing sand. It possesses high strength.

5. **Parting Sand:** It is clean clay-free silica sand. It is used to stop the clinging of green sand to the pattern and also to permit the sand on the surface of cope and drag to detach without sticking.

## Properties of Moulding Sand

1. **Porosity or Permeability:** It is characteristic of moulding sand, which allows passing the gases through it.
2. **Cohesiveness:** It is about the strength of the moulding sand by which it can hold sand particles together.
3. **Plasticity:** It allows all the sand to flow through each of the corners without leaving any space.
4. **Adhesiveness:** The characteristic of the moulding sand is to stick to the surface of the moulding box.
5. **Collapsibility:** The property in which after the solidification, sand mould breaks automatically.
6. **Green Strength:** It is the toughness of sand in humid conditions.
7. **Refractoriness:** The characteristic of the sand is that it withstands high temperatures without getting fused.

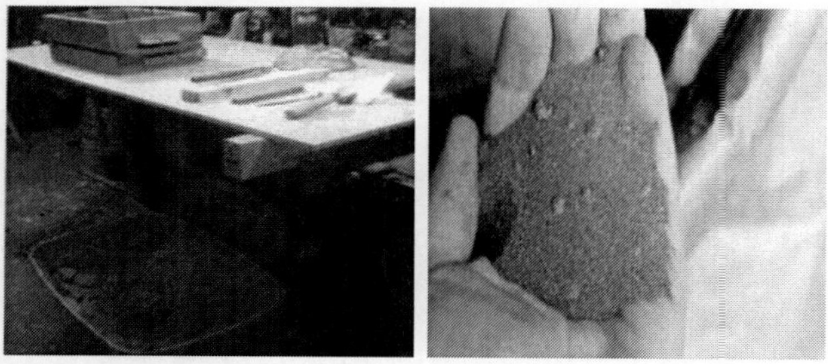

**Figure 205.** Green Sand Mould

## Pattern

The pattern is the reproduction of an item is to be made with a few amendments. These amendments are:

1. Allowances
2. Core print

A pattern is defined as a replica of the object needed to be cast and is embedded in moulding sand around which the pattern has to construct. For generating the cavity, the pattern is then withdrawn in the moulding sand. It contains the projection, which is known as core prints. It is made with some modifications like allowance and core print.

- A pattern that is made properly and has smooth and finished surfaces reduces the defects in casting.

- A pattern that is constructed properly also reduces the overall cost in casting.
- It can be used for the purpose of checking the dimensions of casting.
- It helps in locating the points on the mould.

### Its function

Patterns may be manufactured from softwood such as pine.

They have to now no longer take in any moisture from the moulding sand and consequently, the surfaces of those styles are painted and covered with a water- proof varnish.

A piece of the wooden sample may be used for manufacturing twenty to thirty percent mould a plastic sample may be used for twenty thousand moulds, and a metal sample may be used for up to ninety thousand moulds, relying upon the metallic of the sample.

### Pattern Material

The materials that can be used for making the pattern are plastic, wood, plaster, wax, metal, or mercury.

1. **Wood:** It is a common pattern material because working on it is very easy and convenient.
2. **Plastic:** It is used where metal and wood cannot be useful.
3. **Metal:** It lasts much longer than wood but is costly.

### Selection of Pattern Material

1. According to the shape and size of the casting.
2. According to the material used.
3. According to the number of casting.
4. According to the moulding methods.

### The Alternative of Pattern Materials

1. The range of castings to be produced. Metal Patterns are desired whilst the manufacturing is large.
2. Method of moulding i.e., Hand or gadget moulding.
3. Shape, complexity and length of the casting.
4. Type of moulding substances i.e., Sand etc.
5. The possibilities of repeat orders.

## Types of Pattern

1. **One-Piece or Solid Pattern**: A one-piece or solid pattern is a single solid pattern made of wood, metal, or other materials. It represents the entire object being cast and does not have any movable or detachable parts.
2. **Two-Piece or Split Pattern**: A two-piece or split pattern consists of two parts that are designed to fit together. The pattern is split along a specific plane to allow for easy removal from the mold after casting.
3. **Cope and Drag Pattern**: A cope and drag pattern is used in sand casting, where the cope represents the top half of the mold and the drag represents the bottom half. The pattern is split into two parts, one for each half of the mold.
4. **Three-Piece or Multi-Piece Pattern**: A three-piece or multi-piece pattern consists of three or more parts that are assembled together to form the complete pattern. This type of pattern is used when the object being cast has complex shapes or features that cannot be easily achieved with a single pattern.
5. **Loose Piece Pattern**: A loose piece pattern consists of separate pieces that are not permanently attached to each other. These loose pieces are positioned and secured within the mold to create the desired shape and features.
6. **Match Plate Pattern**: A match plate pattern is a two-piece pattern mounted on opposite sides of a plate known as a match plate. The pattern and plate are used together to create both the cope and drag halves of the mold.
7. **Follow Board Pattern**: A follow board pattern is used in molding when the shape or contour of the pattern cannot be easily created using a regular pattern. It involves using a follow board, which is a template or guide, to shape the molding sand around the pattern.
8. **Gated Pattern**: A gated pattern includes additional structures or features that define the gating system for the casting process. These patterns have channels, gates, and risers integrated into the pattern itself, allowing for proper flow and solidification of the molten metal.
9. **Sweep Pattern**: A sweep pattern is a pattern with a smooth continuous shape, typically created by moving a template or pattern board along a path. It is often used for producing curved or contoured surfaces in castings.
10. **Skeleton Pattern:** A skeleton pattern is a lightweight pattern consisting of a framework or skeleton structure. The pattern represents the basic shape and dimensions of the casting, with the voids filled by the molding material.

11. **Segmental or Part Pattern:** A segmental or part pattern is used when the casting has a large size or complex geometry that cannot be accommodated by a single pattern. It involves dividing the pattern into multiple segments or parts that are assembled to create the complete pattern.

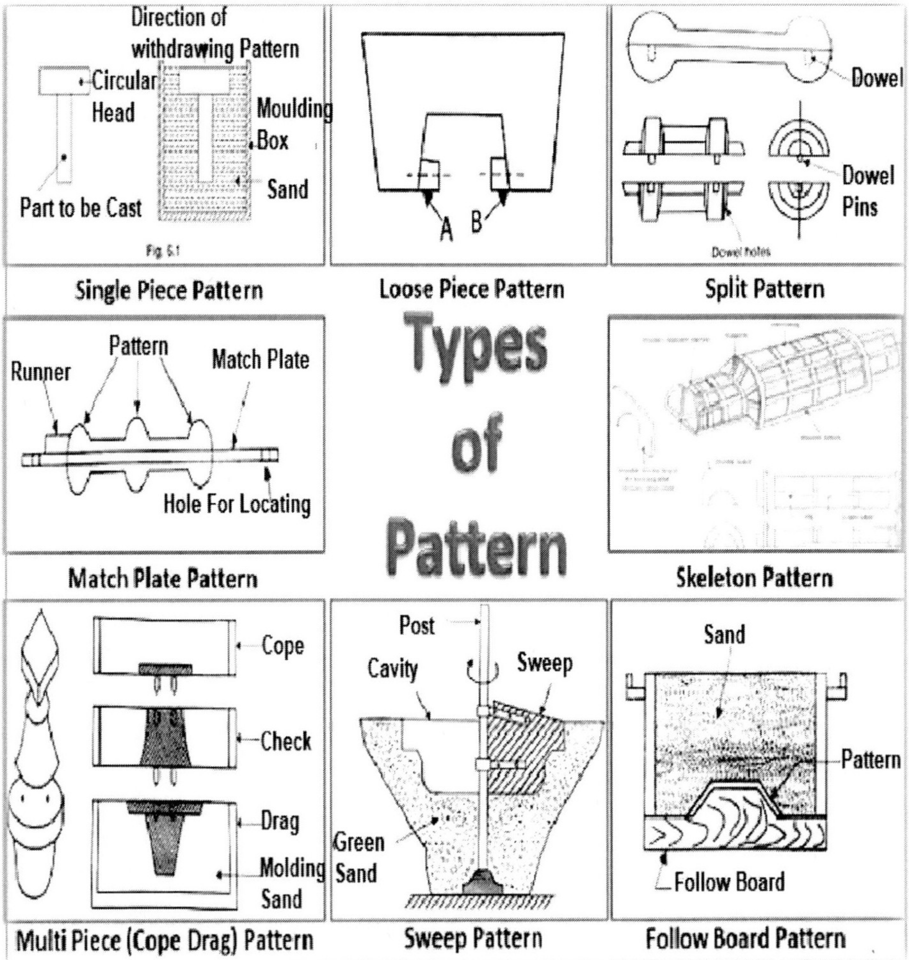


**Figure 206.** Types of Pattern

## Experiment 1

**Aim:** Mould making using one-piece pattern.

**Tools Required**: Trowel, spike, bellows, spruce pin, riser pin, gate cutter, rammers, lifter, molding flask, molding board, slick and shovel.

### Materials Required

- Parting sand.
- Molding sand.
- Dum-bell.

### Sequence of Operations

- To prepare sand.
- To prepare mould.
- Placing of the pattern.
- Preparation of drag and cope.
- To make spure and riser.

### Procedure

1. The drag is first put upside down the moulding.
2. The drag is filled with green sand and it is compacted with the help of a rammer.
3. After ramming, there is a lot of excess amount of sand so to remove excess amount of sand, be used rammer.
4. Now the cope is placed over the drag and they are held together by the means of lugs and pins so that they don't move left or right.
5. After placing the spure pin and riser, we follow the above procedure again.
6. The spure and riser are removed from the cope after proper ramming.
7. Now vent holes are made in mould with the help of the vent wire so that the gases which are released from molten metal in the cavity do not trap and make any defect.
8. The cope and drag are detached and the pattern which is obtained is removed.
9. The runner is made with the help of a gate cutter in the casting.

### Benefits

- It is the cheapest method.
- There is no problem in withdrawing the pattern from drag.
- It is the simplest method.

## Result

Using the given single-piece pattern, the required mould cavity is obtained.

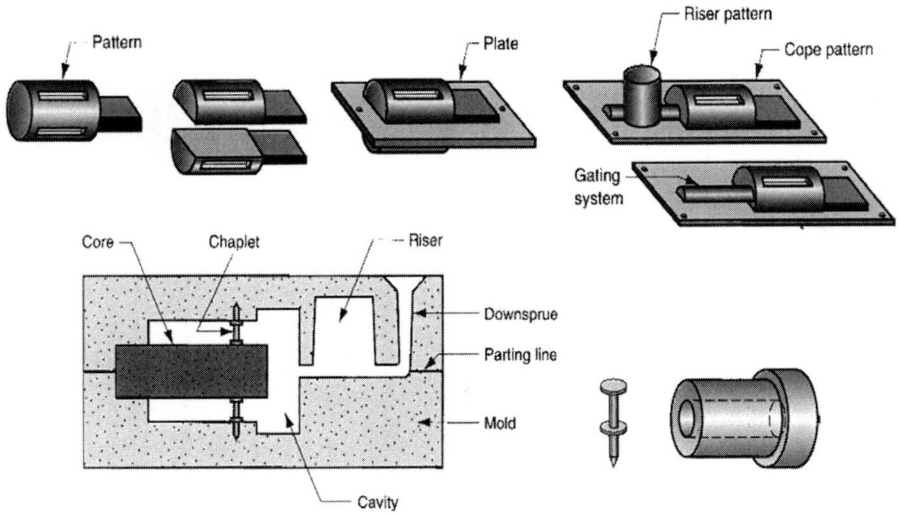


**Figure 207.** Solid Pattern

## Precautions

1. The sand should not be much wet.
2. Ventilation should be provided.
3. The molten metal is too hot so we must be careful while dealing with it.
4. The sand should contain an adequate quantity of moisture.

## Experiment 2

**Aim:** To prepare a mould cavity using a two-piece pattern (split pattern).

**Tools Required:** trowel, spike, gate cutter, lifter, vent rod, molding flask, molding board, slicks, riser pin, spruce pin and bellows.

### Materials Required

- Molding sand
- Parting sand
- Dum-Bell

### Sequence of Operation

- Sand preparation.
- Sand mixing.
- Pouring.
- Preparation of drag.
- Preparation of cope.
- Making of spure and rise.

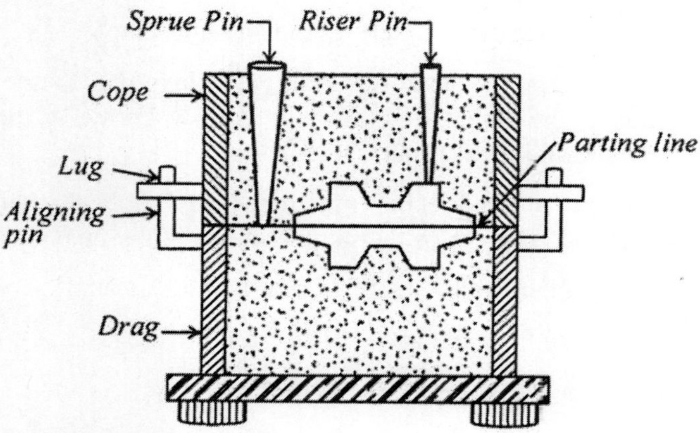


**Figure 208.** Molding Board

### Procedure

1. Place the drag on the molding board, upside down.
2. Sprinkle the pattern and molding board, with parting sand.
3. Allow loose sand, preferably via a riddle over the pattern, unit it's far covered to an intensity of two to three cm.
4. Allow unfastened sand, ideally, via a riddle over the sample, unit its miles blanketed to intensity of two to a few cm.
5. Pack the molding sand across the sample and into the corners of the flask, with fingers.

6. Place a few greater sand within side the flask and % the sample with a rammer, the use of first the peen cease after which butt cease.
7. Strike off the extra sand from the pinnacle floor of the drag with the strike-off bar.
8. Turn the drag upside down.
9. Blow off the unfastened sand debris with the bellows and smoothen the higher Floor.
10. Place the cope directly in the drag in position. Locate the riser pin on the very best factor of the sample.
11. Place the sprue in at approximately five to six cm from the sample on the alternative facet of the riser pin.
12. Sprinkle the higher floor with parting sand.
13. Repeat steps three to 7, approximately.
14. Make holes with the vent rod approximately 1 cm from the sample.
15. Remove the sprue and riser pins by cautiously drawing them out. Funnel formed hollow is made on the pinnacle of the sprue hollow, known as the pouring cup. Lift the cope and area it apart on its edge.
16. Insert the draw pin into the sample. Wet the rims across the sample. Loosen the Sample through rapping. Then draw the sample immediately up.
17. Adjust and restore the mould by including bits of sand, if necessary.
18. Cut gate within side the drag from the sprue to the mould.
19. Blow off any unfastened sand debris in the mould.
20. Close the mould by changing the cope and putting weights on it.

## Result

A sand mould hollow space is ready through the use of Dum-Bell.

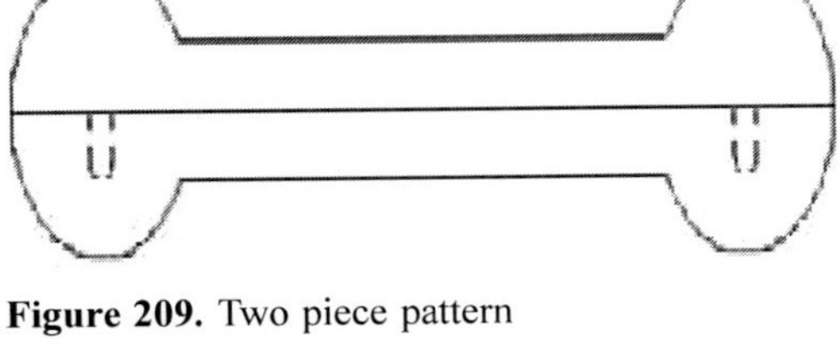

**Figure 209.** Two piece pattern

## Precautions

1. Do not make the sand too wet now.
2. Provide good enough airflow to get rid of smoke and fumes.
3. Never stand close to or appear over the mould all through the pouring due to the molten metal being too hot.
4. Do now no longer shake out a casting too hastily, which might also additionally.
5. Bring about 0.33 degrees.

## Experiment 3

**Aim:** Demonstration of mould-making using a sweep pattern.

### Tools Required

- Base – placed in a sand mould.
- Vertical spindle.
- Wooden template (frame called sweep).
- The outer end of the sweep carries the contour corresponding to the shape of the desired casting.

### Materials Required

- Chaplets
- Mould sand
- Loam sand
- Core
- Green sand
- Dry sand

### Procedure

1. The sweep is rotated about the vertical spindle from the cavity.
2. Then sweep and spindle are removed.
3. Leaving the base in the sand.
4. The hole made by the removal of the spindle is patched up by filling the sand.
5. The separately prepared core is placed in the mould cavity, gates are cut and the mould is ready for pouring.
6. Chaplets are employed to support the core.

### Result

Mould of large symmetrical casting is obtained (circular section).

**Figure 210.** Sweep Pattern

## Precautions

1. Holes are made properly in the mould so that gases are not trapped in the mould.
2. Runners are made properly so that molten metal does not enter the mould directly.

## Benefits

- Large savings in time, labor and material.

**Application:** A sweep pattern is used to make huge kettles.

Sweep Pattern

## Experiment 4

**Aim:** Demonstration of mould using match plate pattern.

### Tools Required

- Moulding box
- Moulding sand.

### Procedure

1. The upper part of the match plate is called the drag.
2. The upper part of the match plate, called the Cope, pattern is put in the cope box and the lower part of the match plate is put in the drag box.
3. Now the mould sand is rammed properly over the match plate.
4. Now the moulding box part is separated so that we match the box pattern and can be able to get out.
5. Now the cope and drag are reassembled and held together with the help of lugs and pins.
6. Now the molten metal is poured into the desired mould so that the component can be cast.
7. Match plates can be made up of magnesium, aluminum and wood, but probably match plates are made up of aluminum because it is not so much costly and it is non-ferrous.
8. In the match plate, the runners and gate are already made.

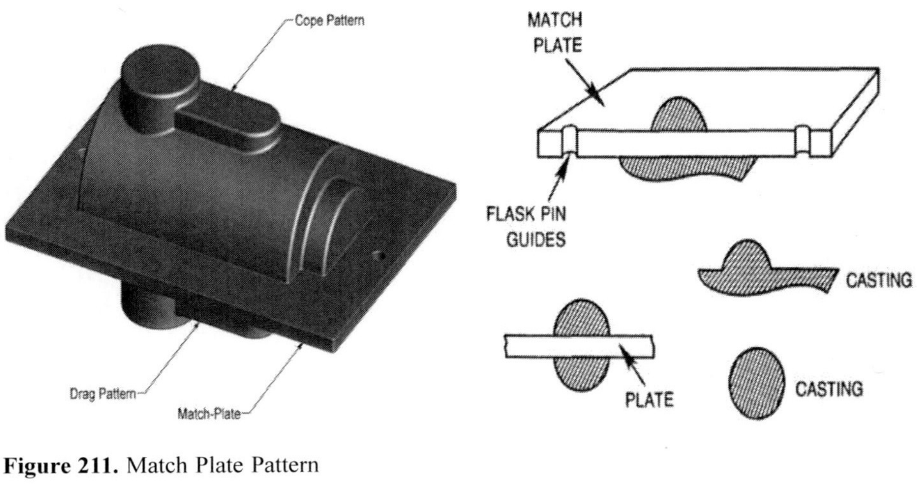


**Figure 211.** Match Plate Pattern

9. The upper part of the match plate is half of the side of a split pattern called the cope pattern.
10. The lower part of the match plate is also half of the side of a split pattern called drag pattern working.

**Result**

Now, with the help of this pattern we can cast the 8 components.

**Precautions**

1. Holes are made properly in the mould , so that gases are not trapped in the mould.
2. Runners are made properly in the mould ,so that molten metal does not get enter directly into the mould.

**Application**

- Through match plate patterns, we can make piston rings and rotor hubs .

**Advantages**

- It is used in higher machine moulding and in mass production.
- Used in higher dimensional control.

## Foundry Shop Objective Questions

1. Which sand has a higher moisture content ?
   a) Dry Sand **b) Green Sand**
   c) Loam d) Facing Sand

2. Which sand has 50% of clay?
   a) Facing Sand **b) Loam Sand**
   c) Dry Sand d) Green Sand

3. Which of the following tool and equipment are used to remove the loose or unwanted sand from the surface?
   a) Lifter b) Ruddle
   c) Swab **d) Bellows**

4. Which of the following characteristic of moulding sand allows them to pass the gas through it?
   a) Plasticity b) Adhesiveness
   c) Green Strength **d) Porosity or Permeability**

5. The property in which after the solidification, sand mould breaks automatically.
   a) Adhesiveness b) Green Strength
   c) Plasticity **d) Collapsibility**

6. Which of the following pattern used to make huge kettles?
   a) One-piece Pattern b) Two-piece Pattern
   **c) Sweep Pattern** d) None of these

7. Which of the following pattern used in higher dimensional control?
   a) Sweep Pattern **b) Match Plate Pattern**
   c) One-piece Pattern d) Two-Piece Pattern

8. Which of the following tools are used for repairing and completing the mold sand?
   a) Ruddle b) Swab
   c) Draw Spike **d) Lifter**

9. Which of the following property allows all the sand to flow through each of the corners without leaving any space?
   a) Green Strength b) Adhesiveness
   c) Porosity **d) Plasticity**

10. The bentonite clay is used inside the walls of the coping box so that.
    a) **Sand does not Stick**
    b) Working Sand becomes Smooth
    c) Working Sand becomes Moisturized
    d) None of the above

## Foundry Shop Subjective Questions

1. What are the most popular pattern-making materials? Discuss their advantages and disadvantages.
2. Classify and describe various tools and equipment used in foundries.
3. What are the common defects of casting? State their causes and remedies.
4. What is pattern making? Explain the split pattern briefly.

# 6

# Machine Shop

## Introduction

A machine shop is a place with machining tools and supplies where hand and power tools are used to make, finish, refurbish, or repair machines and their parts. Machine shops come in helpful for repairing and fabricating a wide range of metal things.

## Machining

Machining is the most extensive technological procedure in the manufacturing industry. The term "machining" refers to the process of cutting out undesired material from raw material to produce the required elements of a product. In this process, manufacturing is accomplished by taking away material (removal of material), it is also known as 'subtractive manufacturing.' Several operations or processes take place in a planned sequence to achieve the desired objective.

## Types of Machines Available in A Machine Shop

1. **Milling Machine :** Solid materials are machined using a milling machine by rotation of the cutting tool.

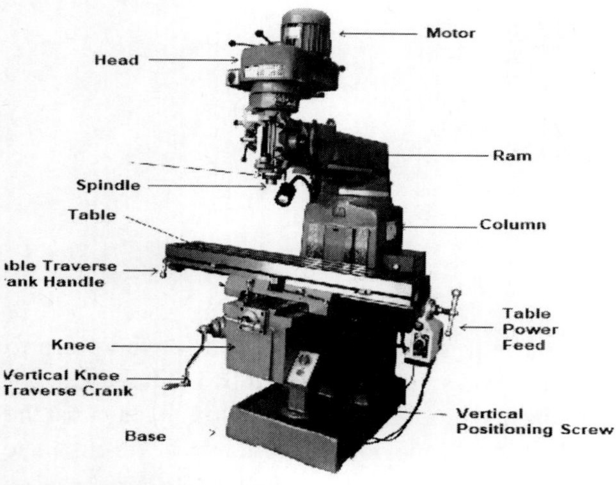


**Figure 212.** Milling Machine

2. **Grinding Machine:** It uses an abrasive wheel as the cutting tool for grinding.

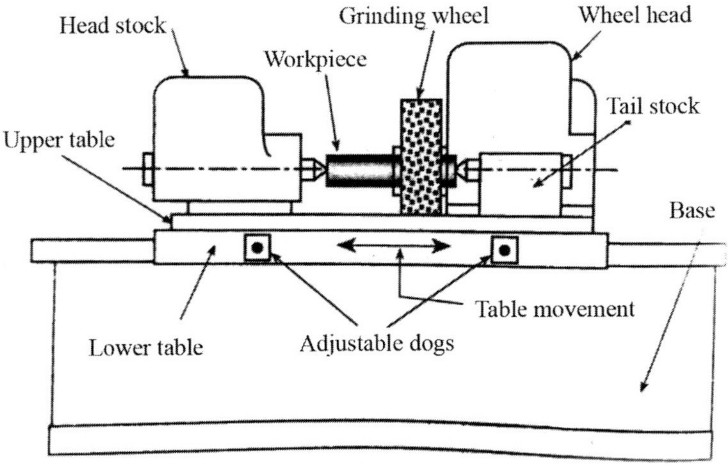


**Figure 213.** Grinding Machine

3. **Planer:** It is used to machine a straight tool path by moving the single-point cut tool and work piece in a linear or straight direction.

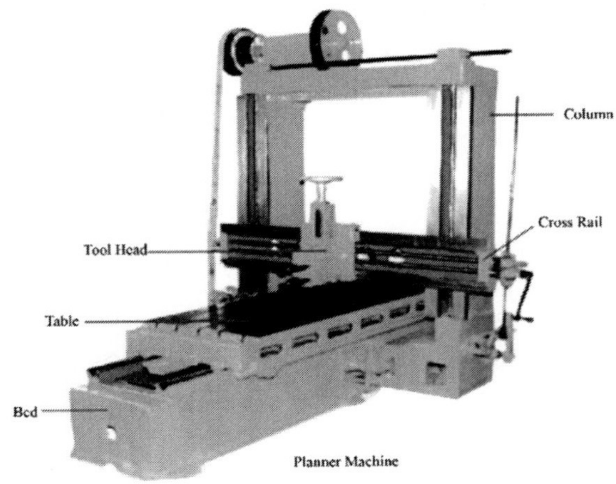


**Figure 214.** Planer Machine

4. **Lathe:** Its primary operation is to eliminate metal from a workpiece to provide it with a specific shape and size. It is finished by holding the work safely and unbendingly on the machine and afterward turning it against the cutting device, which stays fixed on the moving tool post, which will eliminate metal from the work as chips.

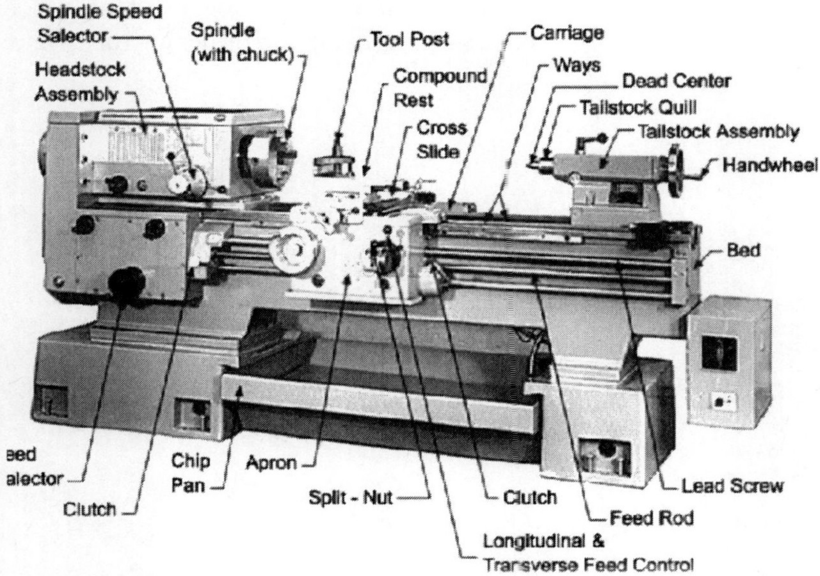


**Figure 215.** Lathe Machine

5. **Shaper:** It is a reciprocating kind of machine in which the ram moves the cutting apparatus, which stays held in a machine to reverse and advances in an orderly forward component. It is utilized to create a level surface through a solitary point cutting device like a mechanical instrument. This is smaller than a planer and the cutter moves in it.

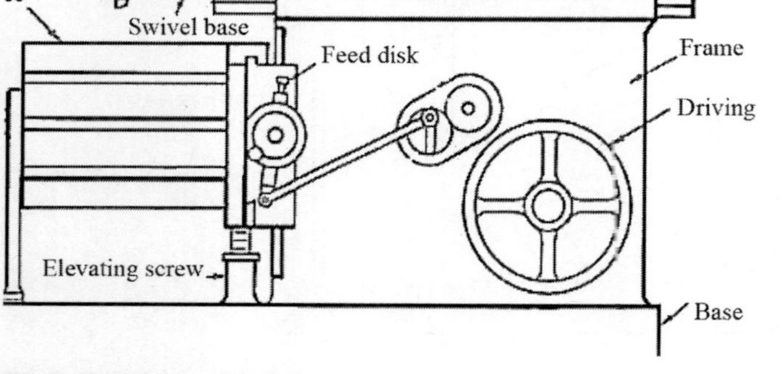


**Figure 216.** Shaper Machine

6. **Broaching Machine:** To remove material for machining, this tool uses a broach or a toothed tool. Two types of broaching can be done – rotary and linear.

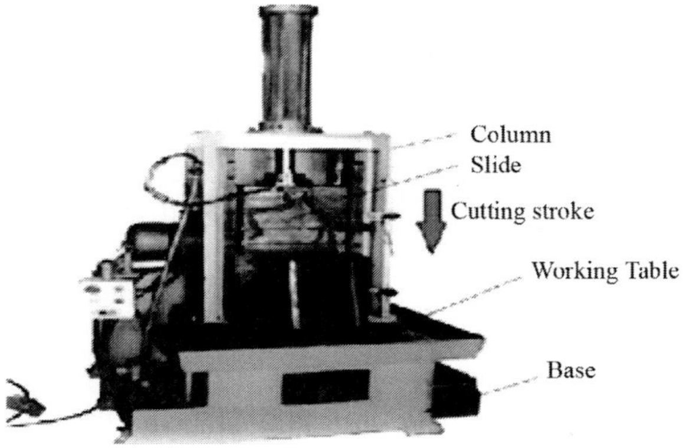


**Figure 217.** Broaching Machine

7. **Slotting Machine:** It is a reciprocating device tool in which the tool reciprocates and cuts while hitting a vertical axis downward.

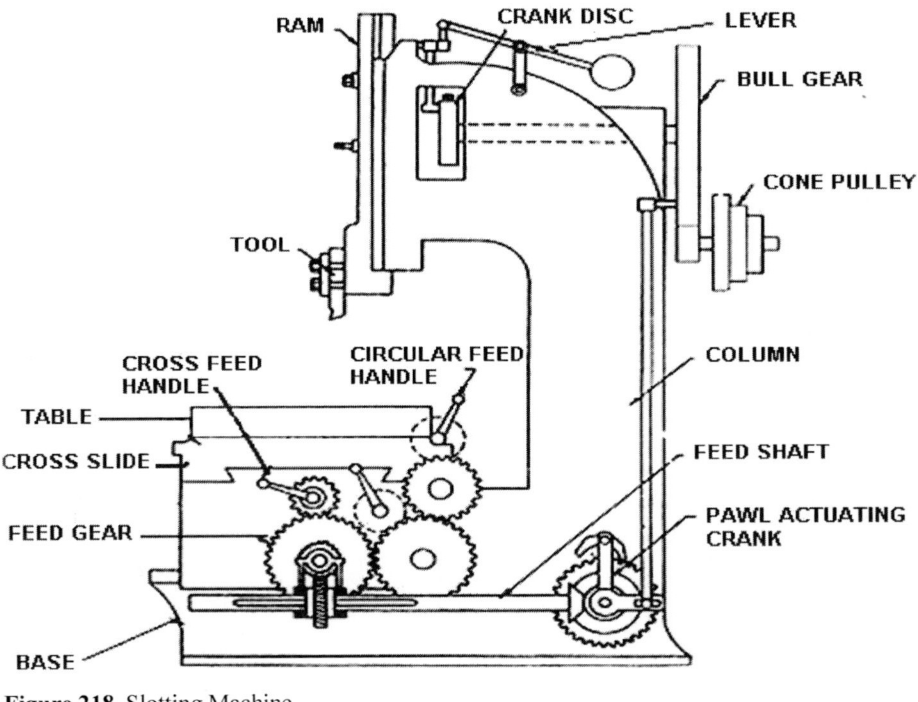


**Figure 218.** Slotting Machine

8. **Hydraulic Press**: A hydraulic press is a machine that uses hydraulic force to exert pressure on objects. It consists of a piston and cylinder arrangement, where a fluid, usually oil, is used to transmit and amplify force. The main purpose of a hydraulic press is to generate a compressive force for various industrial applications such as bending, shaping, crushing, and pressing.

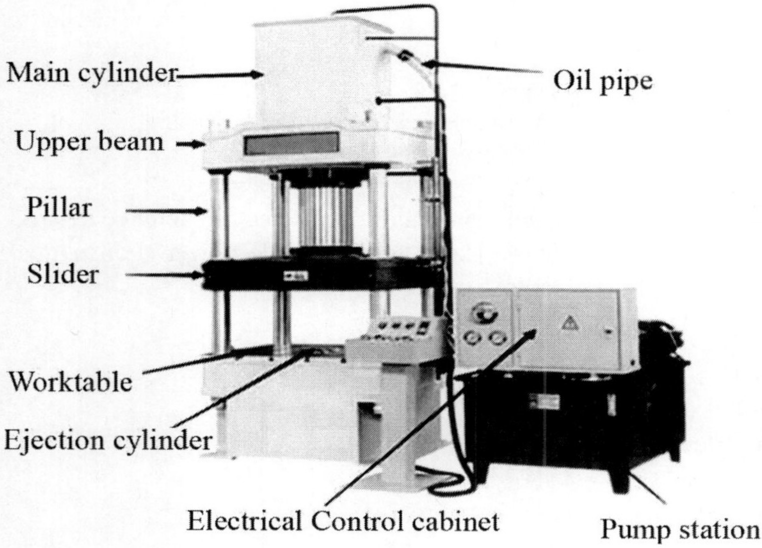


**Figure 219.** Hydraulic Press

## Machine Tools

A machine tool is a device that converts energy into works of the desired machining and finishes by eliminating surplus material from formed blanks using cutting tools that are moved around the work surface.

Tools are equipment that cuts, shears, punches, presses, drills, rolls, grinds, sands, or forms metal, plastic, or wood stock. Tools can be automatic or semi-automatic.

| Machine tools shape metals by: | Machine tools are capable of: |
|---|---|
| 1. Taking off chips. | 1. Holding workpiece. |
| 2. Pressing, drawing or shearing. | 2. Holding cutting tool. |
| 3. Under controlled machining process. | 3. Providing movement to the cutting tool. |
| | 4. Feeding the cutting tool. |

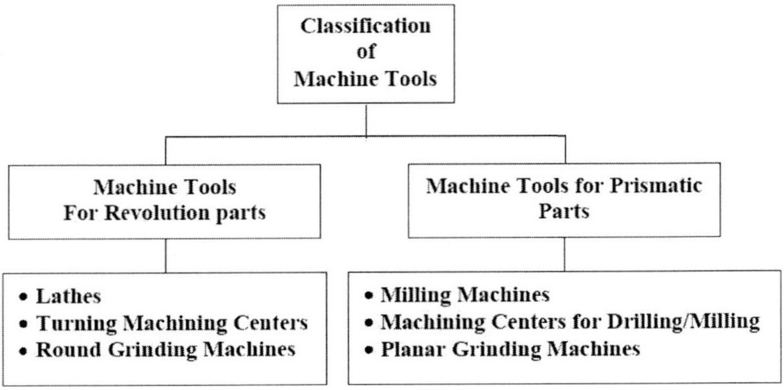


**Cutting Tools:** A cutting tool is an equipment that is used to achieve desired work during the machining process. Hand-held, portable powered or manual tools are all examples of these tools.

## Tool Materials

1. High Carbon steels.
2. High-Speed steels.
3. Cemented Carbides.
4. Medium Alloy steel.
5. Abrasives.
6. Diamonds.
7. Stellite Ceramics.

| | Cutting speed m/min | | | |
|---|---|---|---|---|
| **Work piece material** | **High carbon steel** | **High speed steel** | **Stellite** | **Cemented carbide** |
| Low carbon steel | 24 | 38 | 90 | 140 |
| Medium carbon steel | 20 | 30 | 70 | 115 |
| High carbon steel | 14 | 20 | 50 | 80 |
| Cast iron | 16 | 24 | 60 | 90 |
| Stainless steel | 8 | 12 | 30 | 50 |
| Brass | 45 | 75 | 180 | 300 |
| Bronze | 22 | 35 | 80 | 130 |
| Aluminium alloy | 55 | 90 | 200 | 350 |

## Characteristics of an Effective Tools

1. It must maintain high hardness even at elevated temperatures.
2. It must be wear-resistant.

3. It must resist bulk deformation.
4. It must have a consistent tool life.
5. It must have required thermal properties.
6. The geometry of the tool must be adequate for an operation.
7. It must be chemically stable.

## Types of Cutting Tools

1. **Single Point Cutting Tool:** It has only one cutting edge for main cutting, allowing for simultaneous machining (material removal) in a single pass.

Single-point cutting tools are referred to by a variety of terms

i) **Shank:** The shank is the tool's main body.

ii) **Nose:** The cutting edges are formed by this portion.

iii) **Face:** It is the top surface of a tool on which chips get hit and are removed from the work piece.

iv) **Side:** It is the nearly vertical surface that, together with the bit's end, creates the bit's profile. It is the most prominent surface.

v) **Base:** This is the shank's bottom surface.

vi) **End:** The cutter bit's end is a near-vertical surface that, combined with the side, defines the bit's profile. The trailing surface is what it's called.

vii) **Heel:** The heel of the cutter bit is the part of the base that lies immediately below and supports the face.

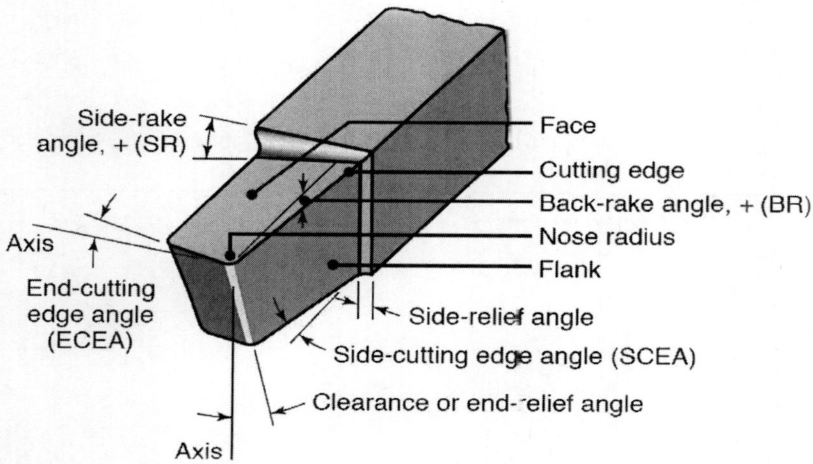


**Figure 220.** Tool Signature

2. **Multi-Point Cutting Tool:** It consists of multiple edges that can perform the machining action. It's a grouping of two or more cutting components (chip- producing devices) attached to a single body.

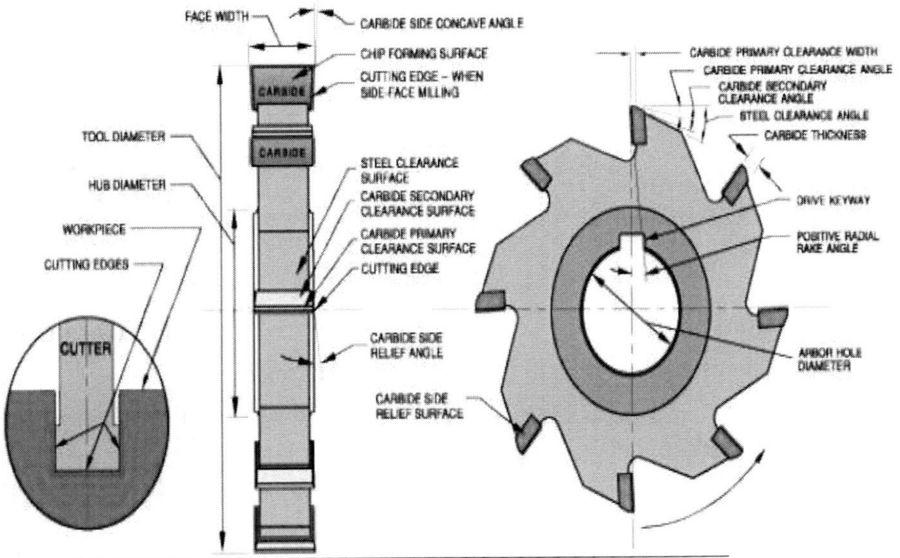


**Figure 221.** Parts and Omportant Angles in Multi-point Cutting Tool

## Types of Cutting Methods

| Orthogonal metal cutting | Oblique metal cutting |
|---|---|
| 1. The cutting edge stays perpendicular to straight the tool's travel direction. | 1. The cutting edge stays inclined at a less than angle to the tool's travel direction. |
| 2. The coiling of the chips takes place in sideways a tight flat spiral. | 2. The chip does not coil but flows in a long curl |
| 3. Sharp corners are produced. | 3. A chamfer is produced at the end of the cut. |
| 4. Life of tool is less. | 4. Life of tools is more. |
| 5. The cutting edge of a smaller length is with the in contact with the work. | 5. The cutting edge of greater length is in contact work. |

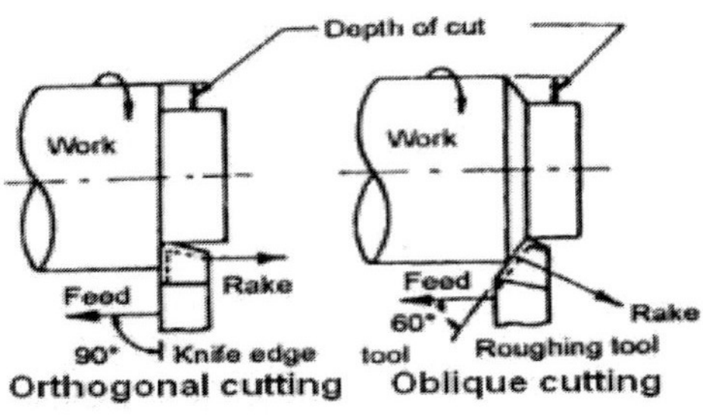


**Figure 222.** Orthogonal and Oblique Cutting

## Cutting Fluid

Cutting fluids are available for a variety of purposes in the cutting process such as to get low deformation, high smoothness and better finish.

## Types of Cutting Fluids Categories

1. Cutting Oils:
   i) Straight oils.
   ii) Compounded.
2. Water Base Cutting Fluids:
   i) Soluble oils.
   ii) Chemical (Synthetic) fluids.
   iii) Semi-Chemical (Semi- Synthetic) fluids.
3. Gases.
4. Paste & Solid Lubricants.

## Functions

1. Reduce friction between tool & material.
2. Reduce the temperature of the cutting zone.
3. Wash away chips.
4. Improve surface finish.
5. Reduce power required.
6. Increase tool life.
7. Prevent welding of the chip to the tool.
8. Prevent corrosion.

## Demonstratioon of Processes in Maching

**Machining Processes**: Machining is one of the processes of manufacturing in which the desired shape of the work piece is provided by removing unwanted extra material. Conventionally this unwanted extra material from the work piece is taken away in the form of chips by feeding the work piece with an appropriate tool. This process can be carried out by single-point or multi-point tools or by abrasive operations. These are summarized below:

1. Single-point tool operations
   i) Turning    ii) Boring
   iii) Planing.    iv) Shaping

2. Multi-point tool operations
   i) Milling.
   ii) Drilling.
   iii) Tapping.
   iv) Reaming.
   v) Hobbing.
   vi) Broaching.
   vii) Sawing.
3. Abrasive operations
   i) Grinding.
   ii) Lapping.
   iii) Honing.
   iv) Super-finishing.

## Demonstration of Processes in Machining

1. **Turning:** Removal of material from the outside diameter of a work piece for proper finishing.

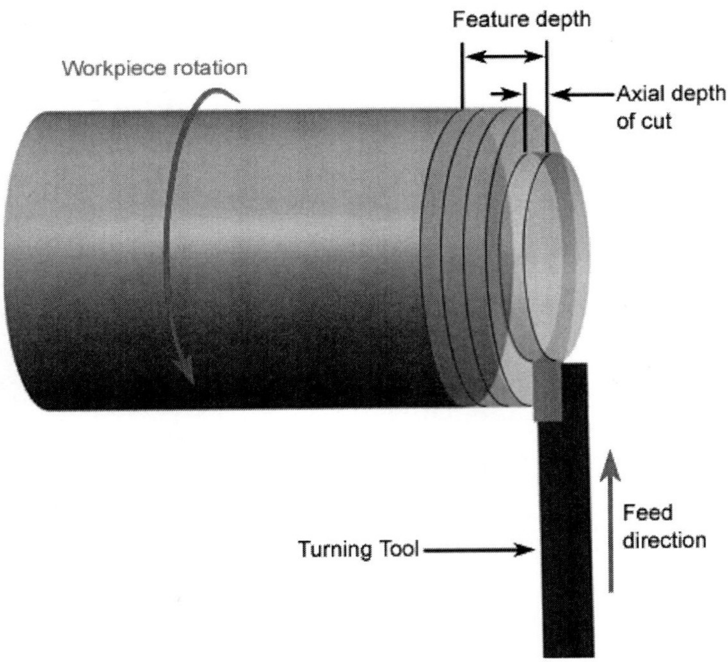


**Figure 223.** Turning Operation

2. **Boring:** In the boring process, already drilled holes can be enlarged or cylindrical cavities can be machined. Internal grooves can also be machined using a boring process.

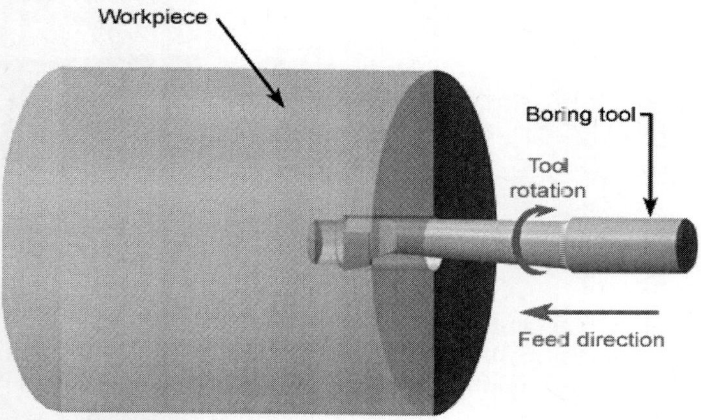


**Figure 224.** Boring Operation

3. **Shaping**: To generate flat, smooth, and contoured surfaces, the tool is repeatedly reciprocated as the work piece is gradually fed into the tool's motion.
4. **Planing:** It is similar to shaping operation but in planning, the work piece reciprocates at a high speed and the tool is fed to the work piece.

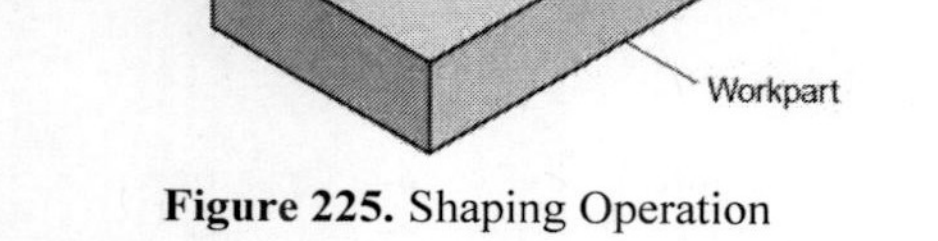


**Figure 225.** Shaping Operation

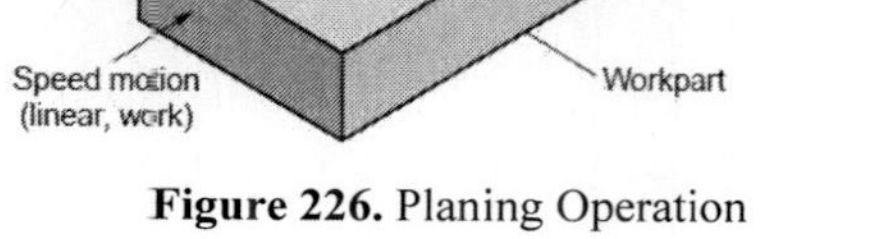


**Figure 226.** Planing Operation

5. **Milling:** In milling operation, the cutting of metal is done with the help of a rotating cutter tool.
6. **Drilling:** By drilling operation, holes are produced on the workpiece with the help of a drill bit.

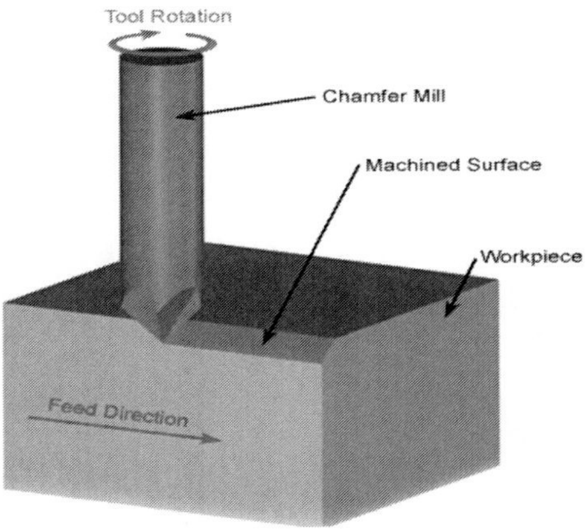


**Figure 227.** Milling Operation

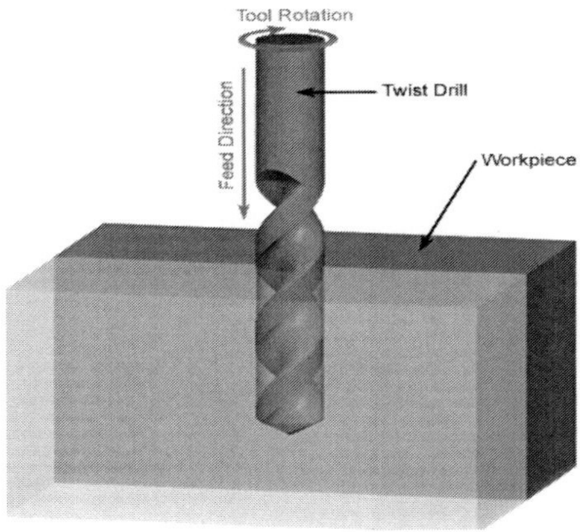


**Figure 228.** Drilling Operation

7. **Tapping:** With the help of tapping operation, threads are made for the purpose of using cap screws, nuts and bolts.
8. **Reaming:** The finishing of the drilled hole is done with the help of a reamer. It provides a surface finish to the drilled holes.

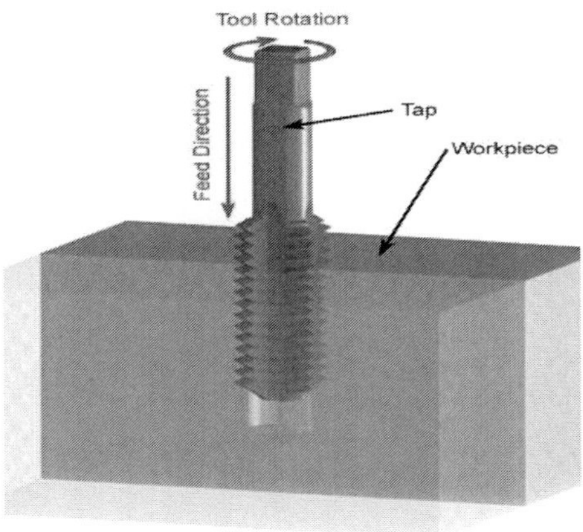


**Figure 229.** Tapping Operation

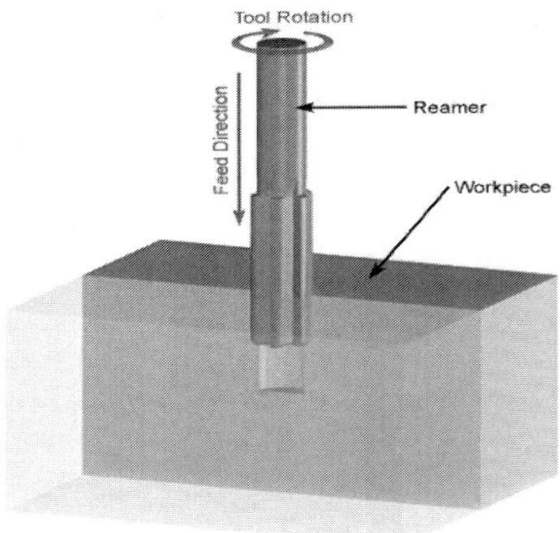


**Figure 230.** Reaming Opeartion

9. **Broaching:** The material is removed continuously and accurately with the help of a toothed tool.
10. **Sawing:** It is the cutting of workpieces with the help of a saw blade having chisel-type teeth.

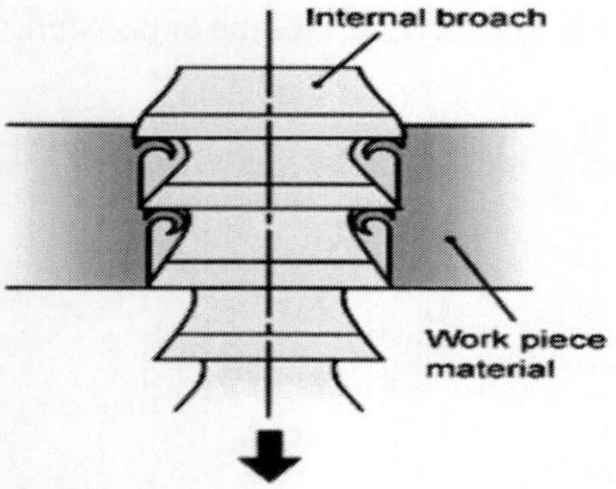


**Figure 231.** Broaching Operation

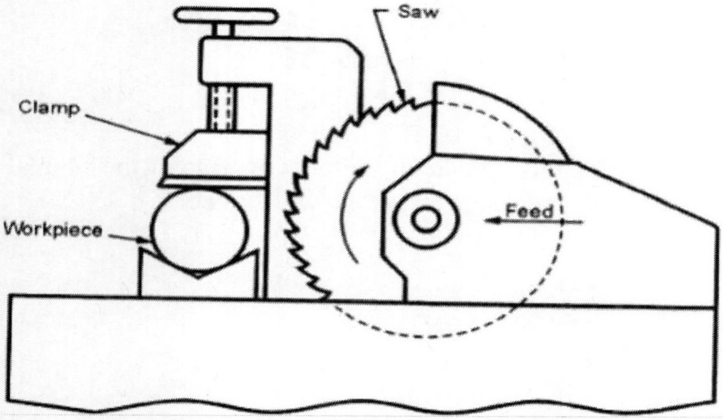


**Figure 232.** Sawing Operation

11. **Grinding:** In this process, the metal is removed with the help of an abrasive wheel (Abrasive machining).

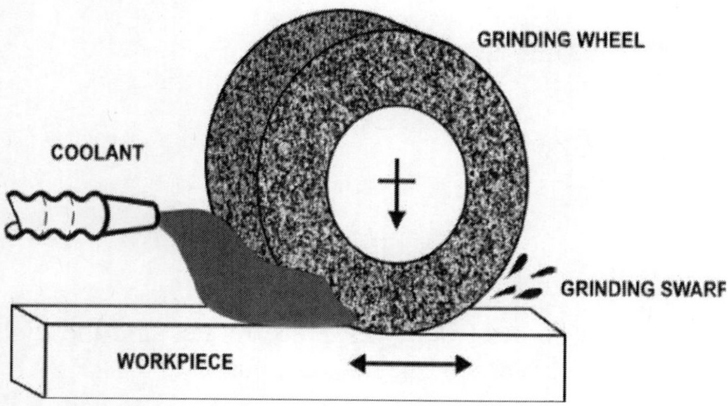


**Figure 233.** Grinding Operation

12. **Lapping:** This process is mainly used for the purpose of sanding or polishing the metal surface for a better finish.

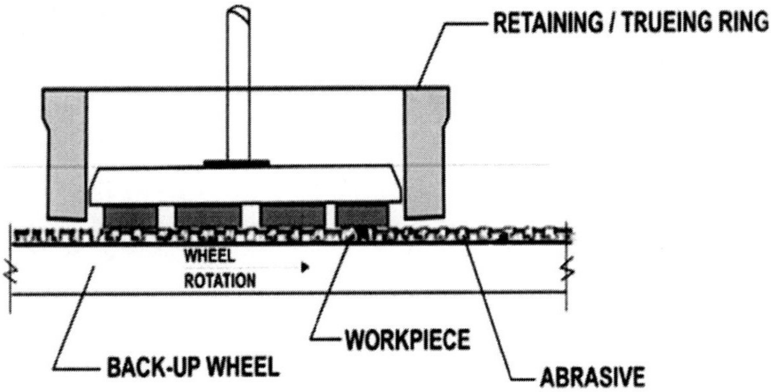


**Figure 234.** Lapping Operation

13. **Honing:** This process is mainly used for the purpose of grinding the internal surfaces for precise finish and shape inside the work piece.

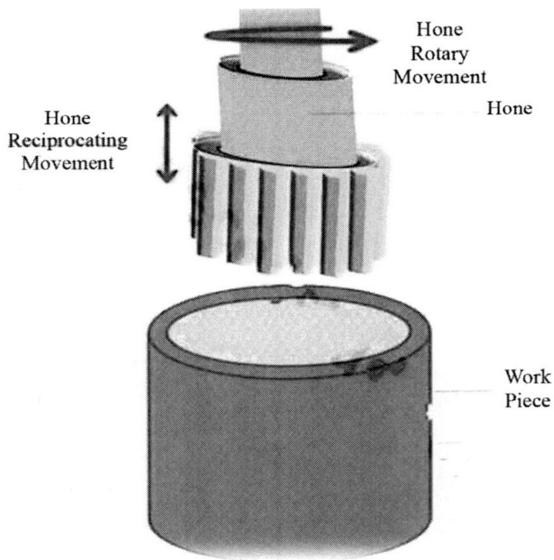


**Figure 235.** Honing Operation

14. **Super-Finishing:** In this process, the work piece rotates and the tool oscillates parallel to the work piece for the finishing purpose.

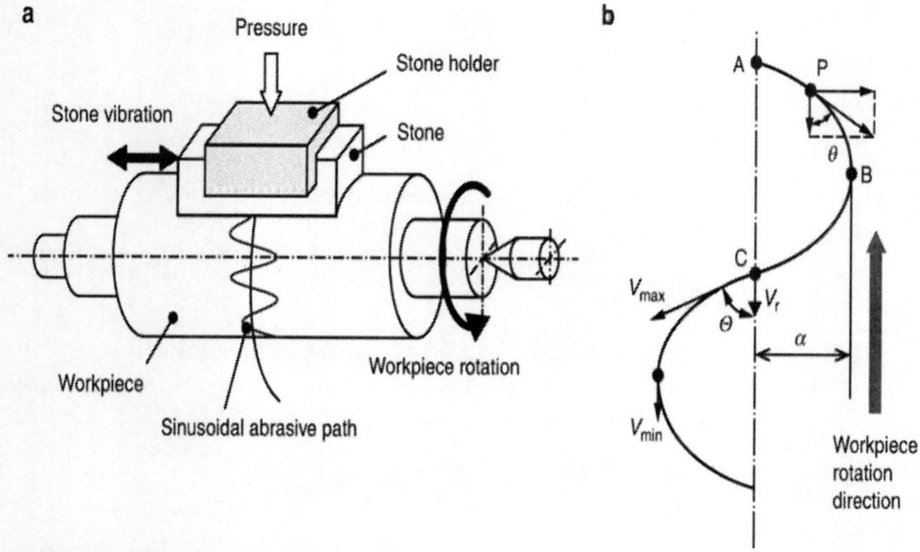


**Figure 236.** Super-finishing Operation

## Uuse of Various Measurement Tools in The Machine Shop

## Methods of Measurement

1. Direct method
2. Indirect method

**Direct method:** The quantity is compared to the primary or secondary standard in the direct comparison technique of measurement. In this case, we're directly comparing the quantity to be measured to the standard.

### Instruments that are used

i) Steel Rule.
ii) Vernier Caliper.
iii) Micrometer.
iv) Measuring Tape.

**Indirect method:** There are a variety of quantities that cannot be directly measured using an instrument. For example, we can't easily measure the strain in the bar caused by an applied force. We may need to take temperature and pressure readings in the deepest parts of the earth or in far-flung locations. In such instances, indirect measurement methods are applied. Instruments that are used:

i) Inside Caliper.

ii) Divider.

iii) Outside Caliper.

iv) Odd leg Caliper.

**Steel Rule:** "A simple measuring tool made out of a long, thin metal strip with a marked unit division scale."

## Types of Steel Rule

1. **Hook Rule :** A hook rule has a small hook or tab at the end, allowing it to be hooked onto edges or surfaces for easier measuring. It is commonly used in woodworking and construction.

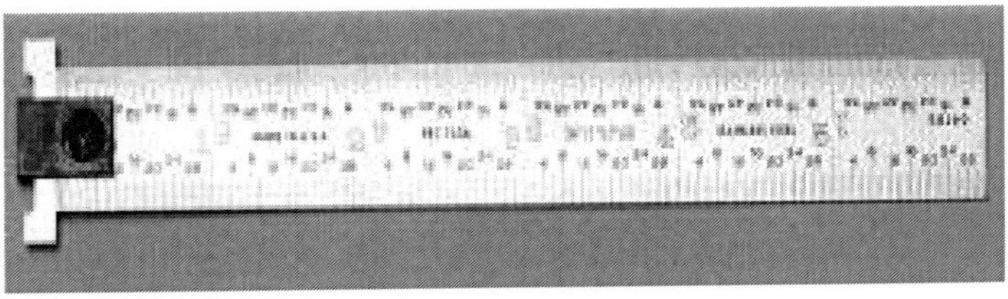

**Figure 237.** Hook Rule

2. **Narrow Rule**: A narrow rule, also known as a narrow steel rule or precision rule, is a type of measuring tool that is characterized by its slim and slender design. It is typically made of high-quality stainless steel and is used for precise measurements in various industries and applications.

**Figure 238.** Narrow Rule

3. **Flexible Rule**: A flexible rule, also known as a flexible steel rule or a measuring tape, is a type of measuring tool that is thin, flat, and flexible. It is typically made of stainless steel or other materials that allow for bending and flexing without losing accuracy. Flexible rules are commonly used in various industries and applications for measuring lengths, widths, and curves.

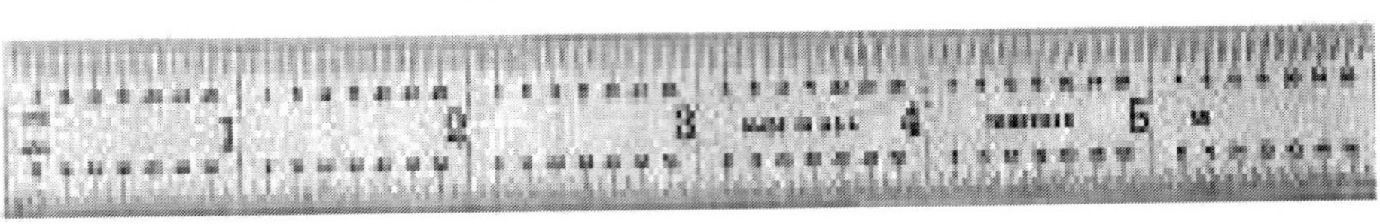

**Figure 239.** Flexible Rule

4. **Narrow Rule with Holder**: A narrow rule with holder refers to a specialized measuring tool that consists of a narrow steel rule attached to a handle or holder for enhanced convenience and ease of use. The holder provides a stable grip and allows for precise measurements without the need to directly handle the rule itself. This combination of a narrow rule and a holder offers increased accuracy and control in various measurement applications.

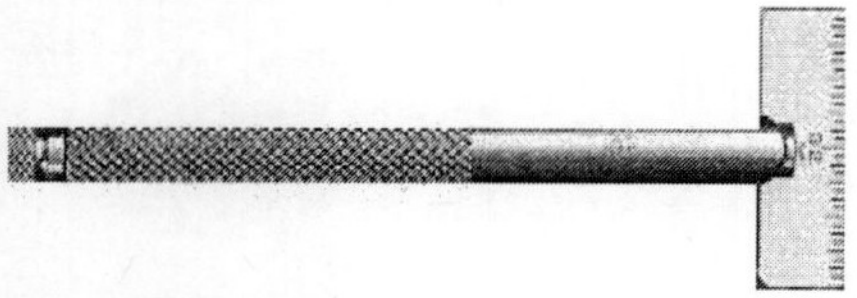

**Figure 240.** Narrow Rule with Holder

## Care of Steel Rule

Steel rules are measuring tools with a high degree of accuracy. Use your steel rule as a scraper, screwdriver, or pry bar only when absolutely necessary. Don't toss it around or drop it. Keep your steel rule gently lubricated at all times. Inspect your steel rule on a regular basis. Make that it isn't bent or dented. Examine the corners for squareness and sharpness. On the steel rule, make sure there are no burrs. Replace your steel rule if you see any of these issues.

**Material:** Stainless Steel

**Sizes:** 6", 12", 18", 24", 3ft and 1 meter.

**Vernier Calipers:** For extremely precise measurements, use a caliper with a vernier scale. It has an L-shaped frame as shown in the figure.

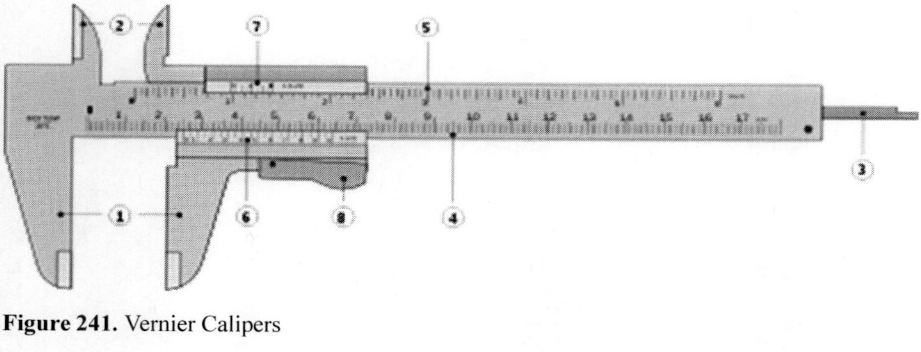

**Figure 241.** Vernier Calipers

## Parts of a Vernier Calipers

1. Outside jaws.
2. Inside jaws.

3. Depth probe.
4. Main scale.
5. Vernier scale.
6. Retainer.

## Least Count

The maximum degree of measurement precision that may be obtained is the least count.

$$\text{Least count of Vernier caliper} = \frac{\text{Smallest reading on main scale}}{\text{No. of division on Vernier scale}}$$

$$\text{Least count of Vernier caliper} = \frac{1\,\text{mm}}{10\,\text{dividions}}$$

$$\text{Least count of Vernier caliper} = 0.1\,\text{mm}$$

$$\text{Least count of Vernier caliper} = 0.1\,\text{mm}$$

**Material:** Stainless steel

**Sizes:** 6 Inch, 8 Inch, 12 Inch, 18 Inch, 24 Inch, and 36 Inch.

## Types of Caliper

1. Dial caliper.
2. Digital caliper.

## Uses

1. To inspect the internal diameter of the work piece.
2. To check the outer diameter of the work piece.
3. To check the length and width of the work piece.

## Holding the Caliper

**Figure 242.** Holding the Caliper

## Precautions in Using Vernier Calipers

1. The instrument must be clean and make sure it works.
2. Check the zero error.
3. Check the parallax error.

**Micrometer:** It is u-shaped. It is a gauge that can measure small distances. It has a threaded spindle which moves slowly toward a small anvil.

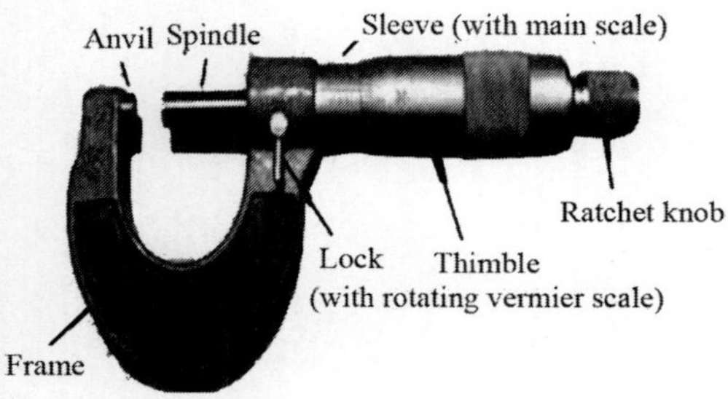


**Figure 243.** Micrometer

**Least Count** = pitch of screw/number of divisions present on a circular scale.

Sizes: 0 to 25 mm, 25 to 50 mm, 50 to 75 mm, 75 to 100, 100 to 125 mm.

**Figure 244.** Least Count

## Types of Micrometer

1. Outside micrometer:
2. Inside Micrometer:

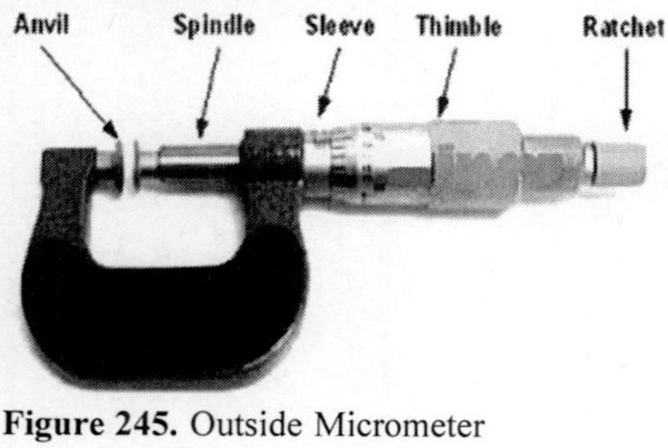


**Figure 245.** Outside Micrometer

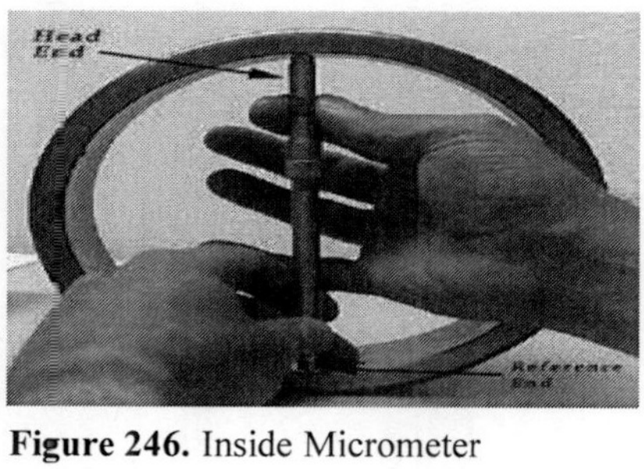


**Figure 246.** Inside Micrometer

3. Depth micrometer:

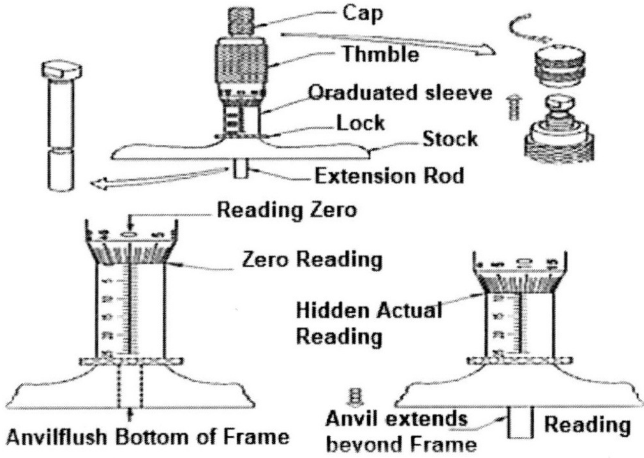


**Figure 247.** Depth Micrometer

4. Bore micrometer:

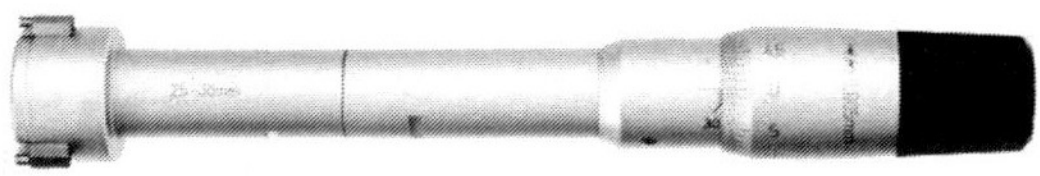

**Figure 248.** Bore Micrometer

5. Tube micrometer:

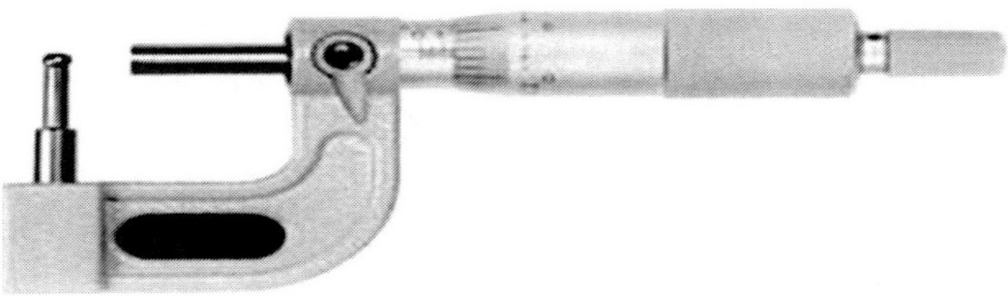

**Figure 249.** Tube Micrometer

## Precautions in using micrometer

1. Check the zero error.
2. Read the measurements correctly.

## Machine Shop Objective Questions

1. Which cutting tool geometry is most suitable for removing material in heavy machining operations?

   a) Positive rake angle  b) Zero rake angle
   c) **Negative rake angle**  d) Neutral rake angle

2. In metal cutting, the process of removing excess material by a multiedge cutting tool is known as:

   a) Drilling  b) Milling
   c) **Turning**  d) Grinding

3. Which machining process is best suited for creating a three-dimensional shape with high precision and complex features?

   a) Turning  b) Drilling
   c) **Milling**  d) Boring

4. Which of the following materials is commonly used as a coolant/lubricant in metal cutting processes?

   a) Water  b) Alcohol
   c) Acetone  d) **Oil**

5. The accuracy of a machine tool is commonly measured by its:

   a) Horsepower  b) Spindle speed
   c) **Tolerance**  d) Surface finish

6. The term "feeds" in machining refers to the:

   a) Rotational speed of the workpiece
   **b) Linear speed of the cutting tool**
   c) Depth of cut
   d) Width of cut

7. Which type of machine tool is specifically designed for producing internal cylindrical surfaces?

   a) Lathe  b) Milling machine
   c) Surface grinder  d) **Boring machine**

8. The process of heat treatment that involves heating a metal to a specific temperature and then cooling it slowly in the furnace is called:

   a) **Annealing**  b) Hardening
   c) Tempering  d) Quenching

9. The micrometer is a precision instrument used for measuring:

   a) **Length** b) Mass

   c) Temperature d) Pressure

10. In which machining process does the workpiece rotate while the cutting tool is stationary?

    a) Milling b) Drilling

    c) **Turning** d) Grinding

## Machine Shop Subjective Questions

1. Define "feed rate" in the context of machining. How does the feed rate influence the material removal rate and the quality of the machined surface?
2. Briefly explain the term "chip formation" during metal cutting. How does chip formation affect the machining process?
3. What is the purpose of using a "center drill" in machining? How does it differ from a regular twist drill?
4. Describe the function of a "tailstock" in a lathe machine. What types of operations can be performed using the tailstock?
5. Explain the purpose of using a "counterbore" tool. When and why would you use a counterbore in machining operations?

## Machine Shop Solved Numericals

1. Calculate the cutting speed in meters per minute for a lathe operation. The lathe spindle is rotating at 800 revolutions per minute (RPM), and the workpiece diameter is 200 mm.

   **Solution:**

   Cutting Speed (V) = $\pi$* Diameter * RPM

   = $\pi$ * 0.2 m * 800 RPM

   = 502.65 m/min

2. Determine the feed rate in millimeters per revolution for a milling operation. The table is moving at a rate of 1500 mm/min, and the spindle is rotating at 600 RPM.

   **Solution:**

   Feed Rate (f) = Table Movement / Spindle RPM

   = 1500 mm/min / 600 RPM = 2.5 mm/rev

3. Calculate the time required to turn a cylindrical workpiece to a final diameter of 100 mm from an initial diameter of 150 mm. The spindle rotates at 300 RPM, and the feed rate is 0.2 mm/rev.

   **Solution:**

   Material Removal = Initial Diameter - Final Diameter
   = 150 mm - 100 mm
   = 50 mm

   Time (t) = Material Removal / (Feed Rate * Spindle RPM)
   = 50 mm / (0.2 mm/rev * 300 RPM)
   = 833.33 seconds (or 13.89 minutes)

4. A workpiece needs to be turned to a final diameter of 80 mm from an initial diameter of 150 mm. The tool can only remove 0.5 mm per pass. Calculate the number of passes required and the total time if each pass takes 2 minutes. The spindle rotates at 250 RPM.

   **Solution:**

   Number of Passes
   = (Initial Diameter - Final Diameter) / Material Removal per Pass
   = (150 mm - 80 mm) / 0.5 mm/pass
   = 140 passes

   Total Time = Number of Passes * Time per Pass
   = 140 passes * 2 minutes/pass
   = 280 minutes

5. Convert a cutting speed of 120 meters per minute to revolutions per minute (RPM). The workpiece diameter is 0.3 meters.

   **Solution:**

   RPM = Cutting Speed / (π * Diameter)
   = 120 m/min / (ð * 0.3 m)
   H" 127.32 RPM

## Machine Shop Unsolved Numericals

1. A machine shop is facing the task of drilling a hole through a steel plate that is 25 mm thick. If the drill bit feed rate is 0.1 mm/rev and the spindle speed is 800 RPM, calculate the total number of revolutions required to drill through the plate.

   **Ans.:** Total Revolutions = 250 revolutions.

2. In a machine shop, a lathe is used to turn a steel rod with a diameter of 40 mm. If the cutting speed is set to 200 m/min and the workpiece is rotated at 500 RPM, calculate the required feed rate for the lathe tool.

   **Ans.:** Feed Rate = 200 m/min / (π × 40 mm) H" 1.5915 mm/rev.

3. A machine shop is using milling to machine a flat surface on a steel block. The block is 150 mm long and 80 mm wide. If the milling cutter has a feed rate of 0.2 mm/tooth and there are 6 teeth on the cutter, calculate the time required to complete the milling operation.

   **Ans.:** Time = 150 mm / (0.2 mm/tooth × 6 teeth) = 125 minutes.

4. In a machine shop, a cylindrical component is being turned on a lathe. The initial diameter of the component is 100 mm, and the final diameter after turning is 80 mm. If the tool feed rate is 0.15 mm/rev and the length of the component is 200 mm, calculate the total number of revolutions required for turning.

   **Ans.:** Total Revolutions = 133.33 revolutions.

5. A machine shop is performing drilling operations on a metal workpiece. The workpiece is rotated at 1200 RPM, and the drill bit feed rate is set to 0.08 mm/rev. If the diameter of the hole drilled is 15 mm, calculate the cutting speed.

   **Ans.:** Cutting Speed = π × 15 mm × 1200 RPM H" 56549.33 mm/min

# 7

# Lathe Machine

## Introduction

It is considered to be one of the most ancient machine tools, having evolved from the early tree lathe, which was a new mechanism for turning and milling items held between two nearby trees at the time. The job rotated sporadically due to a rope wound around the work, with one end linked to a flexible limb of a tree and the other end being pulled by a worker. Then hand tools were employed.

The machine was given the name "lathe" when a strip of wood called "lath" was used to support the rope during its construction. This technology evolved over centuries, and in 1797, an Englishman named Henry Maudslay invented the first screw-cutting lathe, which is the ancestor of today's high-speed, heavy-duty production lathe.

## Function

A lathe's principal purpose is to remove the metal from a workpiece in order to provide it with a specific form and size. This is done by securing and tightly mounting the work on the machine, then spinning it against a cutting tool that removes metal from the work in the form of chips.

## Working Principle of Lathe Machine

1. A lathe is a machine tool used to remove unwanted materials from a workpiece in the form of chips with the help of a tool that travels across the workpiece and can be fed deep into the work.
2. When the tool is moved parallel to the work-piece, then the cylindrical surface is formed.
3. If the tool is moved inclined to the axis, then it produces a tapered surface and so called taper turning.

The working principle of the lathe machine is shown in figure no. 250

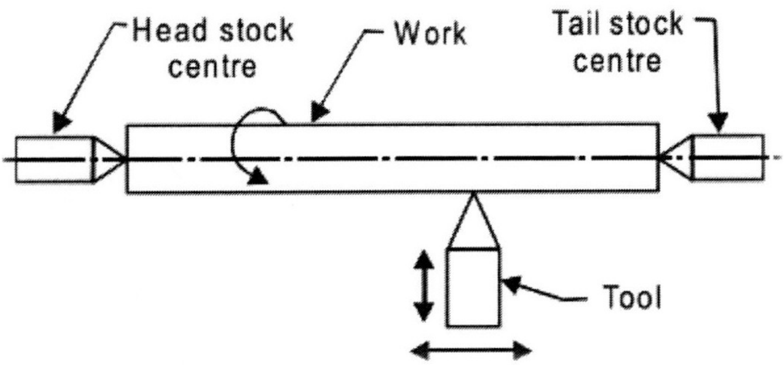


**Figure 250.** Working Principle of Lathe Machine

## Size of Lathe

The maximum size of work that can be handled by the lathe is used to designate the size of the lathe (that is, the diameter and length of the work). Usually, the maximum diameter of the work is specified first, and the length is specified as the maximum distance between centers.

Manufacturers usually use the term swing to designate the size as the maximum diameter of the work that can be machined in the lathe. Thus, a 16-inch lathe indicates that the machine is designed to machine work up to 16 inches in diameter. Manufacturers also use the length of the lathe bed, rather than the distance between centers, to indicate lathe size. The swing of a lathe refers to the nominal swing and not to the actual swing of the lathe. The actual swing is the radial distance from the center axis to the bed and is always slightly greater than the nominal swing.

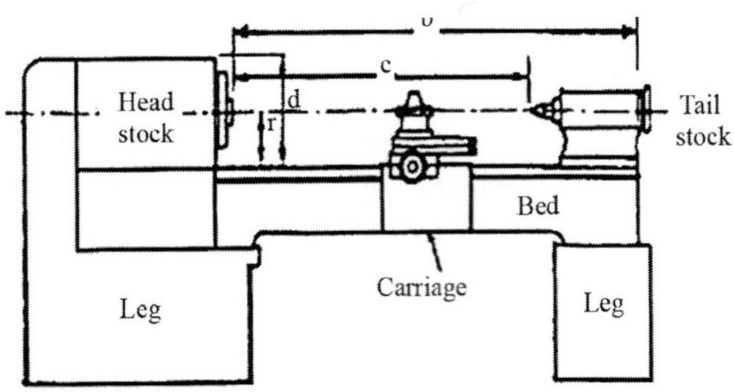


r- Centre height

d- swing diameter over bed

c - Length between centres

b- Length of bed

**Figure 251.** Size of Lathe

The size of a lathe is shown in figure no. 246 is expressed or specified by the following items :

1. The height of the centers was measured from the lathe bed.
2. The swing diameter over the bed. This is the largest diameter of work that will revolve without touching the bed and is twice the height of the centre measured from the bed of the lathe.
3. The length between centers. This is the maximum length of work that can be mounted between the lathe centers.
4. The swing diameter over the carriage. This is the largest diameter of work over the bed.
5. The maximum bar diameter. This is the maximum diameter of the bar stock that will pass through the hole of the headstock spindle.

## Classification of Lathe Machine

1. Speed Lathe.
   i) Wood cutting lathe.
   ii) Centering lathe.
   iii) Polishing lathe.
   iv) Spinning lathe.
2. Engine Lathe.
   i) Belt driven lathe.
   ii) Individual motor drive lathe.
   iii) Gear head drive lathe.
3. Bench Lathe.
4. Tool room Lathe.
5. Capsten and Turret Lathe.
6. Special purpose Lathe.
   i) Wheel lathe.
   ii) Gap-bed lathe.
   iii) T-lathe.
   iv) Duplicating lathe.
7. Automatic Lathe.

1. **Speed Lathe:** It is the simplest type lathe. Due to high speed of spindle it is called "Speed Lathe". Speed is 1200 to 3600 rpm used for polishing and spinning of metal.

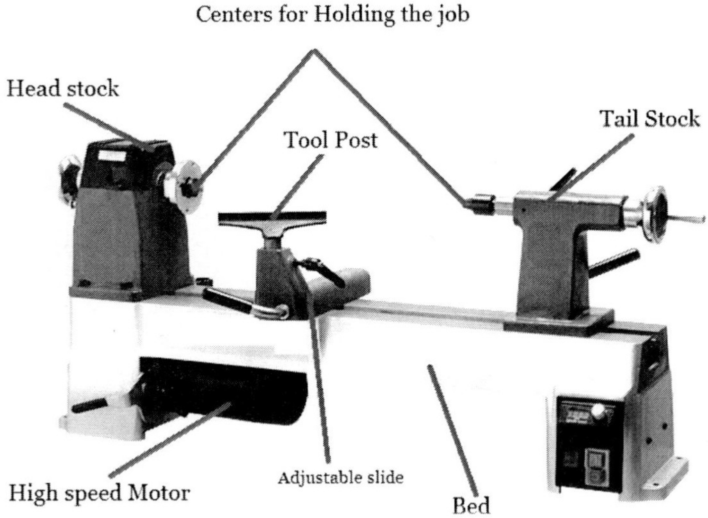


**Figure 252.** Speed Lathe

2. **Centre Lathe:** It is motor driven machine and available in a large variety and shapes. Short runs are preferable in this lathe. Earlier it was run by the steam engine. Hence, it is also called Engine Lathe.

**Types**

i) Belt-driven lathe.

ii) Individual motor-driven lathe.

iii) Gear head drive lathe.

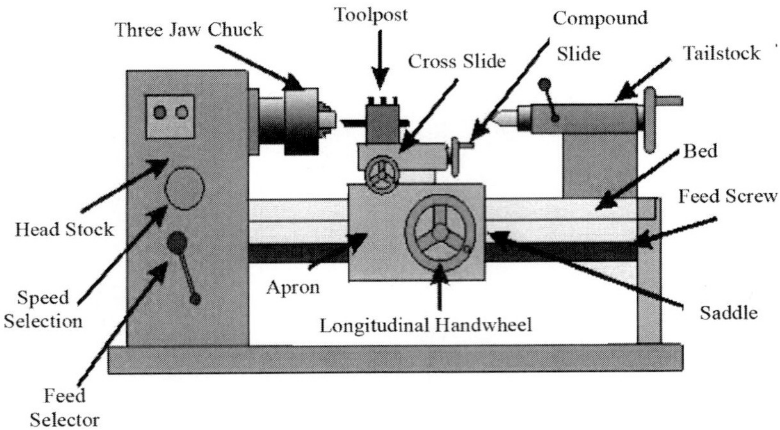


**Figure 253.** Centre Lathe

3. **Bench Lathe:** It is small in size and fitted on the bench. It is used for Small light weight workpieces.
4. **Tool Room Lathe:** Same as engine lathe. It is used for small works like gauges and fixtures, etc. Different speeds can be achieved. Costly as compared to a same-size Engine lathe.

**Figure 254.** Bench Lathe

**Figure 255.** Tool Room Lathe

5. **Capsten and Turret Lathe:** Capsten lathes: for light work pieces. Turret lathes: for heavy work pieces. Does not consist of the tailstock. Tail stock is replaced by the hexagonal turret.
6. **Automatic Lathe:** In this machine, all operations are automatically performed, such as the workpiece fed and removed without the help of an operator.

**Figure 256.** Capsten and Turret Lathe

**Figure 257.** Automatic Lathe

7. **Special Purpose Lathe:** They are programmable and computerized numerically controlled lathes.

**Figure 258.** Special Purpose Lathe

## Description of Lathe Parts

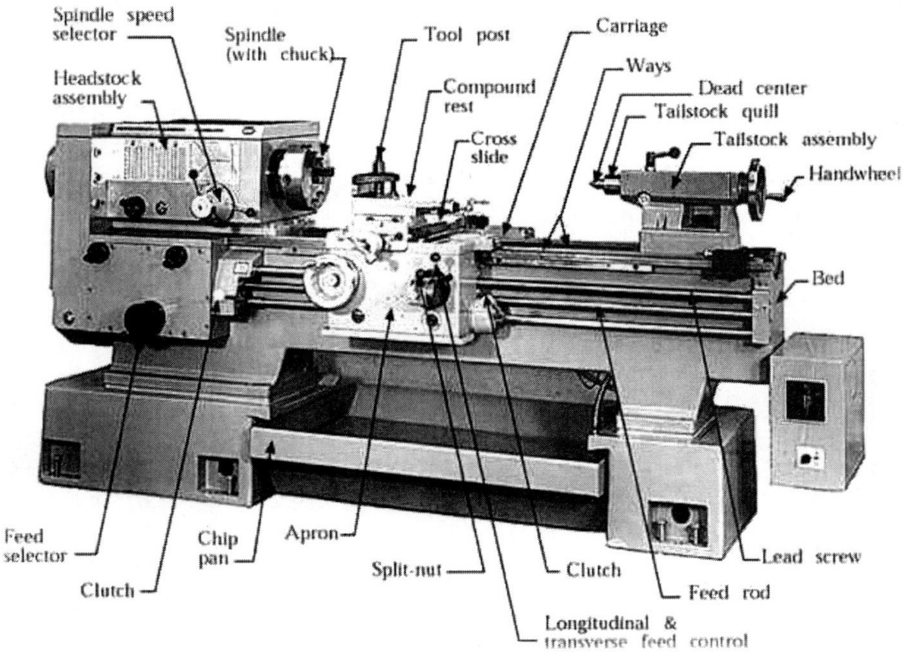


**Figure 259.** Lathe Parts

## Lathe and its Parts

1. Bed.
2. Headstock.
3. Tailstock.
4. Carriage.
5. Feed mechanism.
6. Screw cutting mechanism.

1. **Bed:** The machine's foundation is the lathe bed. The carriage lies over the lathe bed and slides on it, with the headstock and tailstock placed at each end. Cast iron with nickel and chromium alloys is the best material for making the bed.
2. **Headstock:** It is attached to the left end of the bed. It consists primarily of a hollow spindle and a mechanism for driving and controlling spindle speed. The headstock casting houses all of the pieces. Carbon or nickel-chrome steel is suitable for spindles.
3. **Tailstock:** At the right end of the bed, on the inner ways, is the tailstock. This serves two purposes:
   i) Supports the opposite end of the piece.
   ii) As a tool holder for various operations.
4. **Carriage:** A lathe's carriage is made up of many components that support, move, and regulate the cutting tool. It is made up of the following sections: the saddle, the cross-slide, the compound slide or compound rest, the tool post, and the apron.
5. **Saddle:** It is H-shaped casting. It houses the cross slide as well as the tool post.
6. **Cross Slide:** It has locations for the tool post or compound resting on the upper face.
7. **Compound Rest:** The compound rest, also known as a compound slide, is a circular base graduated in degrees that is set on top of the cross-slide.
8. **Tool Post:** This is positioned on the top of the compound rest and is used to hold the tool and modify it to a comfortable operating posture.
9. **Apron:** The apron is attached to the saddle and drapes over the bed's front. It has gears, clutches, and levers for manually and power-feeding the carriage.

**Feed Mechanism:** "Feed" refers to the tool's movement to work. Types of feed on a lathe tool:

1. Longitudinal
2. Cross
3. Angular

**Different Feed Units**

1. End of bed gearing.
2. Feed gear box.
3. Feed rod and lead screw.
4. Apron mechanism.

**Tool Holders:** Used to hold tool bits. Available as:

1. **Right Hand Tool Holder**:- A right-hand tool shown in figure no. 260 removes material when moving from right to left (it takes only the top view keeping the rake surface at the top). The name right-hand tool was used based on the human hand analogy. Using the human right hand, the thumb represents the tool feed direction. Therefore, the principal cutting edge of the tool is on the tool's left side.

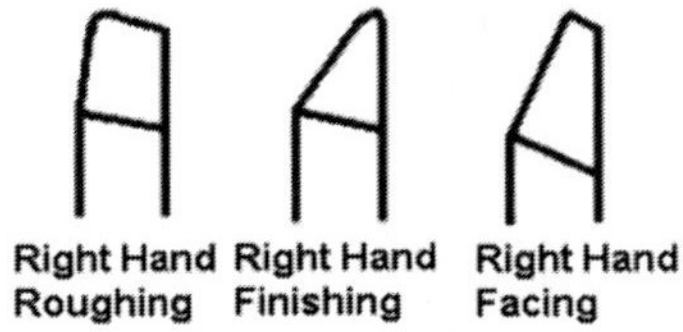


**Figure 260.** Right Hand Tool Holder

2. **Left-Hand Tool Holder** :- Unlike a right-hand tool shown in figure no. 261, the left-hand cutting tool removes material when moving from left to right (top view keeping rake surface visible). Using the left human hand, the thumb represents the tool feed direction. Therefore, the principal cutting edge of the tool is on the tool's right side.

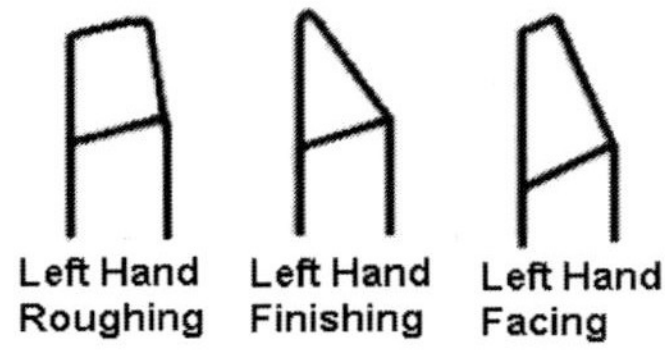


**Figure 261.** Left Hand Tool Holder

3. **Round Nose Tool Holder**:- A round nose turning tool shown in figure no.262 has no back rake or side rake. Therefore, the tool is fed from the left to the right or from the right to the left-hand end of the lathe bed. However, in some cases, the round-nose tool can have a small back rake. Due to the round edge, round nose turning tools are popular tools in finish turning operations.

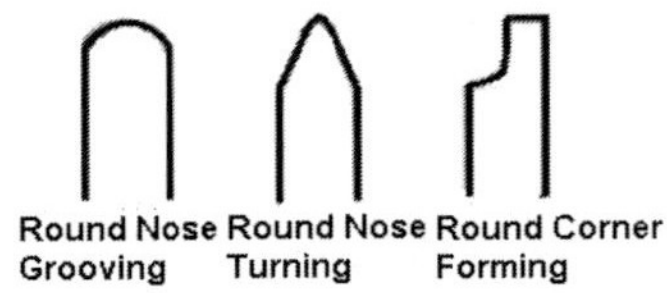


**Figure 262.** Round Nose Tool

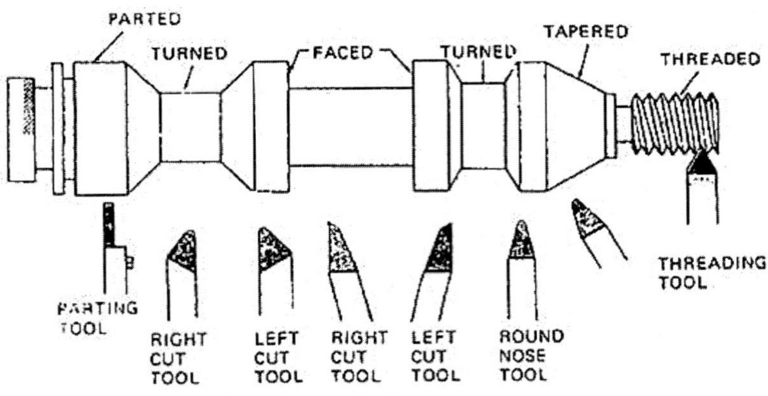


**Figure 263.** Lathe Machine Cutting Tool

**Cutting tools:** Made of high-speed steel or ceramic (tungsten carbide). Shown in figure no. 263 and 264.

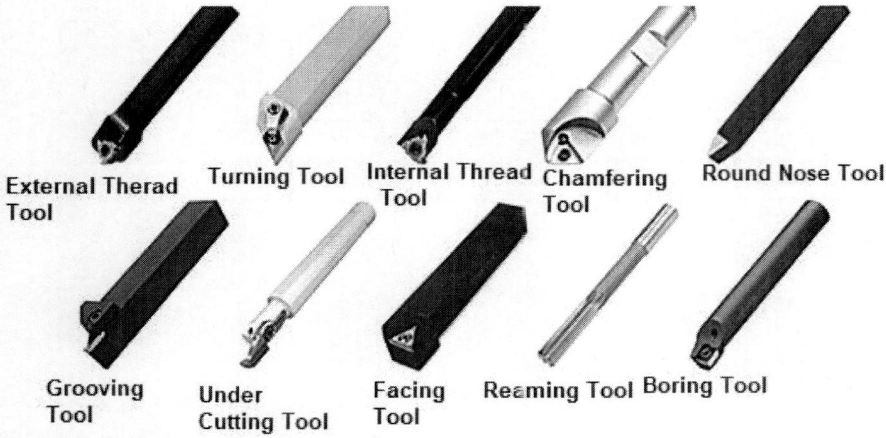


**Figure 264.** Types of Cutting Tool

## Classification of Cutting Tool by Material

### 1. High-Speed Steel Tool

As its name sounds, the lathe machine tool is made of high-speed steel (HSS), which can maintain hardness although it is put in a service temperature of up to 1000° F. (588°C). Therefore, they are ideal for rough machining and semi-finish machining.

### 2. Carbide Tool

The cutting edge of the carbide tool comes from carbide. Therefore, they are ideal for cutting iron, plastic, glass, stone, steel, and other difficult materials.

### 3. Diamond Blade

Diamond is a very hard and wear-resistant material. It has a low coefficient of friction, high elastic modulus, high thermal conductivity, and low thermal expansion coefficient. Consequently, the lathe tools with diamond cutting edge is ideal for precision machining of brittle, wear-resistant, conforming, and hard materials such as graphite.

### 4. Other Materials

The cutting edge of these types of machine lathe tools comes from other materials such as cubic boron nitride and ceramic. Therefore, they are ideal for rough machining and cutting hard alloys.

## Cutting Speed in m/min.

| Material | Turning | Thread Cutting | Drilling | Reaming |
|---|---|---|---|---|
| Cast Iron | 15 – 19 | 7 – 8 | 22 – 31 | 6 – 8 |
| Mild Steel | 25 – 31 | 9 – 10 | 28 – 35 | 10 – 15 |
| Brass | 60 – 90 | 20 – 25 | 60 – 90 | 25 – 30 |
| Aluminium | 120 | 25 – 30 | 60 – 90 | 20 – 30 |

**Figure 265.** Cutting Speed

## Mathematical Form

$$Cs = \left[\frac{\pi DN}{1000}\right]$$

The rotational speed in turning is related to the desired cutting speed at the surface of the cylindrical workpiece by the equation.

$N = v/\pi d$

*Where* N = rotational speed, rev/min; v = cutting speed, m/min (ft./min); and Do= original diameter of the part, m (ft.). The turning operation reduces the diameter of the work from its original diameter Due to a final diameter $D_o$ as determined by the depth of cut d.

$D_f = Do - 2d$

The feed in turning is generally expressed in mm/rev (in/rev). This feed can be converted to a linear travel rate in mm/min (in/min) by the formula.

$F_r = N f$

*Where* f r = feed rate, mm/min (in/min); and f = feed, mm/rev (in/rev). The time to machine from one end of a cylindrical work-part to the other is given by.

$T_m = L/f r$

*Where* Tm = machining time, min; and L = length of the cylindrical work-part, mm (in).

### Cutting Conditions in Turning

A more direct computation of the machining time is provided by the following equation.

$T_m = \pi D o L/f v$

*Where* Do = work diameter, mm (in); L= work-part length, mm (in); f=feed, mm/rev (in/rev); and v = cutting speed, mm/min (in/min). As a practical matter, a small distance is usually. Added to the work-part length at the beginning and end of the piece to allow for approach and over-travel of the tool. The volumetric rate of material removal can be most conveniently determined by the following equation.

The cutting speed is calculated from the rotating speed for any turning operation as the following.

C.S = $\pi$ DN/1000

*Where,*

C.S = Cutting Speed

D = Día of work piece (mm)

N = Speed in RPM (c/min)

Π = Constant

## Lathe Accessories

1. Chucks :
   i) Three jaw chuck.
   ii) Four jaw chuck.
   iii) Collet chuck.
2. Centres:
   i) Live centre.
   ii) Dead centre.
   iii) Tipped centre.
   iv) Ball centre.
3. Mandrels :
   i) Plain mandrel.
   ii) Stepped mandrel.
   iii) Collared mandrel.
   iv) Screwed mandrel.

## Lathe Operations

Operations performed when work piece is held between centres or by a chuck:

1. Straight turning.
2. Shoulder turning.
3. Taper turning.
4. Eccentric turning.
5. Chamfering.

6. Thread cutting.
7. Facing.
8. Knurling.
9. Filling.
10. Polishing.
11. Grooving.
12. Spinning.
13. Spring winding.

When the work piece is held by a chuck/faceplate/angle plate:

1. Drilling.
2. Reaming.
3. Boring.
4. Counter boring.
5. Taper boring.
6. Internal thread cutting.
7. Tapping.
8. Undercutting.
9. Parting off.

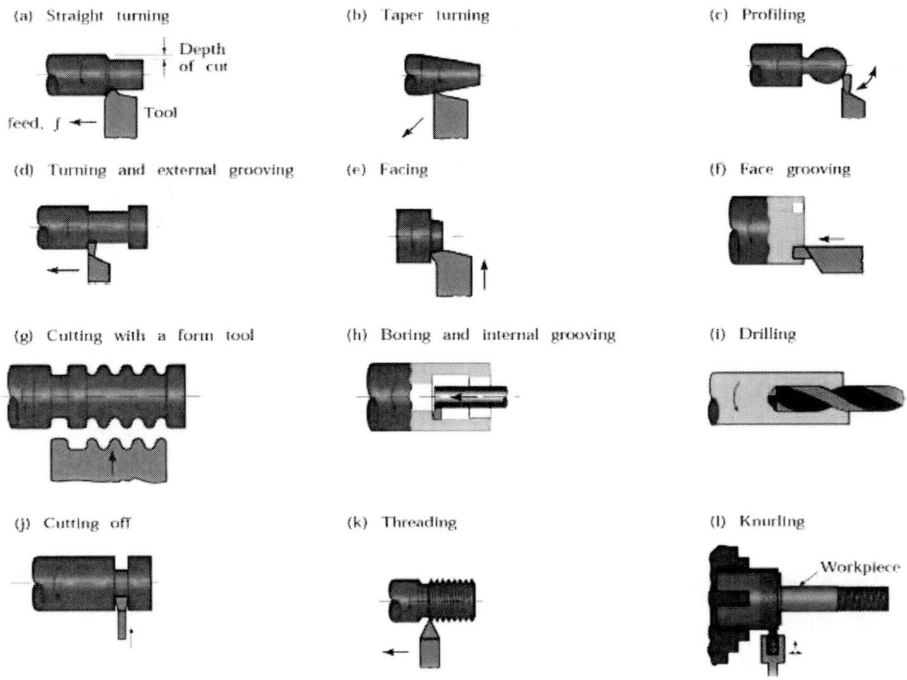


**Figure 266.** Lathe Operations

## Operations performed with special attachments

1. Grinding.
2. Milling.

## CNC Lathe Machine

A lathe is traditionally used in woodturning, metalworking, metal spinning, thermal spraying, parts reclamation, and glass-working. One or two centers typically hold the workpiece, and at least one center can be moved horizontally to accommodate workpieces of different lengths. CNC lathes are quickly replacing older production lathes due to their ease of setting, operation, repeatability, and accuracy. CNC lathe work is just one type of CNC machining.

CNC stands for computer numerical control and refers to a machine operated by a computer system. A CNC lathe is operated with precise design instructions to machine parts to precise specifications. A CNC lathe shown in figure no. 268 is a machine tool where the material or part is held in place and rotated by the main spindle as the cutting tool that works on the material is mounted and moved on various axes.

**Figure 267.** CNC Machine

A simple CNC lathe operates on two axes with the cutting tool in a fixed position at an 8 to 24-station turret. The rotating action of the workpiece is called "turning." This is why certain types of CNC lathes are called CNC Turning Machines. CNC lathes produce precise round shapes with an outer diameter (OD) and an inner diameter (ID). Many different kinds of structures can be machined with this machine tool, depending on their various industries' needs. They are commonly used in oil and gas, automotive, aerospace, medical, electronics, mining, power plant, steel and paper mills, and shipbuilding industries.

Modern CNC lathe machines usually come with tools that can work on multiple axes. Different types of machines can include both X and Y-axis, incorporate a second turret with other tools, or include a second spindle. This means a CNC lathe can perform additional milling, drilling, and tapping operations, resulting in a complete and even complex part manufacturing process and eliminating the need to transfer the part to another machine for a second operation. This article looks at the different types of CNC lathes, their applications, and their considerations.

## Types of CNC Lathe

CNC lathes are classified into types by how many axes they have. They are available with different axes, allowing more complex parts to be produced without manually switching machines or tools. The different axes affect how the machined part or the tool can be positioned, rotated, and approached during the machining process.

### • 2-Axis CNC Lathe

This basic CNC lathe machine has two linear axes that can perform outer diameter/inner diameter, basically cylindrical machining, and facing operations, or drilling and tapping at the center of the part. It has an X and Z axis and does not allow for milling.

### • 3-Axis CNC Lathe

With a 3-axis CNC lathe, a C axis and life tool system are added, allowing the part to be positioned so that standard milling operations, boring, and tapping operations can be performed. Helical milling operations that require the slow sync rotation of the workpiece are possible.

### • 4-Axis CNC Lathe

A 4-axis machine has an added Y-axis so that off-center machining operations can be performed. The fourth axis works well for more irregular and complex lathework.

### • 5-Axis CNC Lathe

In 5-axis lathes, a second turret is added to a 3-axis CNC lathe. Such machines would then have two axes at each upper and lower turret plus the C-axis on the rotating spindle. This enables the use of two tools at the same time on the part, which drastically increases machining speed.

### • 6 Or More Axis CNC Lathe

With more than five axes, the permutation may include the following: two C-axes with a primary and second spindle; two turrets with an upper and lower turret, each with two linear axes; one Y-axis at the upper turret; and a second spindle that can move towards the main spindle to pick up the part. There are even machines with more than eight axes. However, 8-axis lathes are highly sophisticated and complex machines that aren't generally needed for everyday manufacturing needs.

## Applications of CNC Lathes

The material being worked by a CNC lathe machine is slowly sheared away. The result is a precisely finished product or an intricate part. Because these machines are so versatile, many industries use them, including automotive, electronics, aerospace, firearm manufacturing, sporting, and others.

A CNC lathe can produce plane surfaces and screw threads or, in the case of ornate lathes, three-dimensional products that are quite complex. They can be used on both small and large parts with extreme accuracy. The workpiece is typically held firmly in place by one or two centers, but the material can also be secured with collets or clamps.

Some examples of items a CNC lathe machine can make include baseball bats, camshafts, bowls, crankshafts, cue sticks, sign boards, musical instruments, and table and chair legs.

## Experiment 1

**Aim:** To perform various operations (turning, thread cutting, knurling, facing operations) in accordance with the provided drawing.

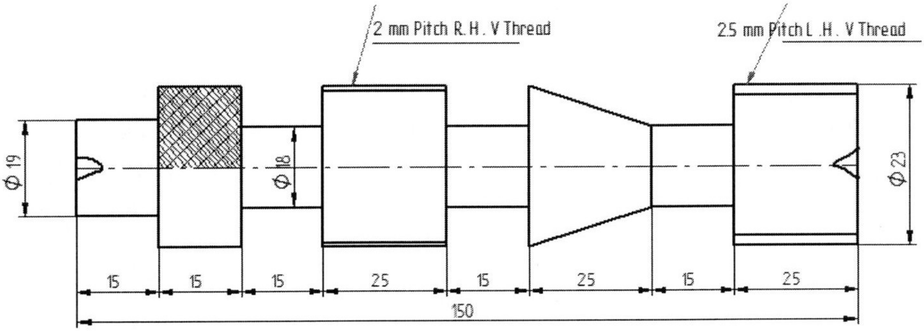


**Figure 268.** All Dimensions Area in mm Turning, Thread Cutting, Knurling, Facing Operation

### Tools Required

- Work piece.
- Vee Tool.
- Surface gauge.
- Surface plate.
- V-Block.
- Centre Punch.
- Lathe Machine.
- Single Point Cutting Tool.
- Drill bit.
- Work piece.
- Chuck key.
- Tool post key.
- Inside and outside Caliper.
- Steel Scale.

### Procedure

1. The diameter of the specified work item is measured with the Outside Calipers.
2. A counter hole is drilled in the center of the work piece after the facing operation is completed on both sides.

3. Insert the workpiece into the three-jaw chuck.
4. The tailstock is moved to retain the work piece in the middle of the dead center and the chuck.
5. Align the tool by placing it such that the tool's tip is perpendicular to the dead center's tip. The align key is used to tighten the tool post.
6. Using a micrometer provided in the compound rest or in the carriage, the depth of the cut is adjusted to achieve a diameter of 23mm.
7. Prior to turning on the machine, cutting parameters such as speed, feed, and depth of cut are selected.
8. For a specified length of 150mm, a plain turning operation is performed.
9. Step turning is performed with the help of markings.
10. Groove turning operations are performed then.
11. The compound rest is swiveled to the angle "theta" in degrees (as determined) using the align key. On each run, the taper turning is done for –

    i) Length = 25 mm

    ii) Depth of cut = 1 mm.

    $$\theta = \tan^{-1}\frac{(D-d)}{2l}$$

    *Where*

    $\theta$ = Swivel the compound rest to the calculated angle.

    D = Major diameter.

    d = Minor diameter.

    l = Length of the taper.
12. A particular knurling tool is used to perform the knurling procedure.
13. The required pitch for thread cutting is noted down, and the depth of cut is computed using a formula. The auto feed varying mechanism is set for 2 mm pitch for right-hand thread cutting operation and a 2.5mm pitch for left-hand thread cutting operation using a standard chart. This is accomplished by engaging a lead screw that rotates at the required speed and direction for the mentioned thread, as well as appropriate gears in the gear train.
14. The technique is done 13 times, with each cut depth incrementing by 0.2 mm divisions. When cutting thread with the right hand, the feed will be from right to left.
15. The technique is done 16 times, with each cut depth incrementing by 0.2 mm divisions. When cutting thread with the left hand, the feed will be from left to right.

16. For 'V' – thread cutting, the thread angle is 600 degrees, and the ISO standard specifies that metric thread depth = 0.65X pitch (60 to 65 percent).
17. Check the precision of the thread pitch with a pitch gauge after turning and the diameter of the turned portion with a Vernier caliper.
18. On the medium feed, a very little depth of cut (say 0.2mm) is used to finish the job.

## Observations

1. Work piece material: ___________
2. Cutting Tool Material: ___________
3. Feed: ___________ mm/revolution.
4. Depth of Cut ___________ mm
5. Selected Cutting Speed ___________ m/min.
6. Plain Turning ___________ m/min
7. Step Turning ___________ m/min
8. Taper Turning ___________ m/min
9. Thread Cutting ___________ m/min
10. Profile Turning ___________ m/min
11. Knurling operation: ___________ m/min

## Result

The required design of the work piece is obtained.

## Precautions

1. Wear a safe outfit.
2. Maintain a safe distance from cutting tools during machining.
3. Work under the supervision of an expert only.
4. Note the readings accurately.
5. Before beginning the experiment, check the machine speed.

## Experiment 2

**Aim:** To perform various operations (Turning, Step-Turning, Facing Operations) in accordance with the provided drawing.

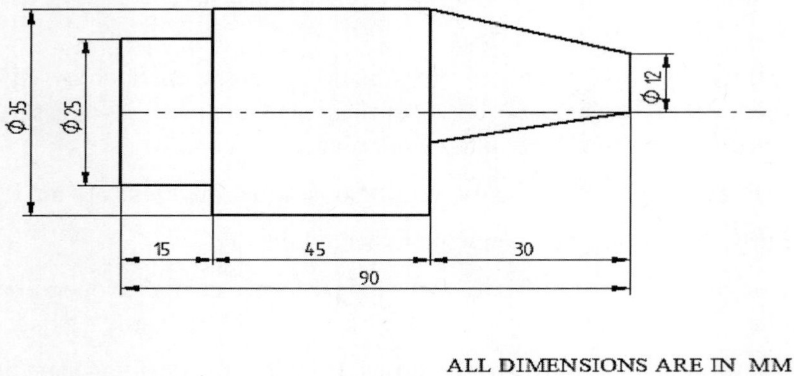


**Figure 269.** Turning, Step-Turning, Facing Operations

## Tools Required

- Work piece.
- Vee Tool.
- Surface gauge.
- Surface plate.
- V-Block.
- Centre Punch.
- Lathe Machine.
- Single Point Cutting Tool.
- Drill bit.
- Work piece.
- Chuck key.
- Tool post key.
- Inside & outside Caliper.
- Steel Scale.

## Procedure

1. The diameter of the specified work item is measured with the Outside Calipers. The work piece is inserted in the four-jaw chuck, and the surface gauge is used to centre it.

2. On both sides of the work piece, the facing operation is performed.
3. The tool is made to touch the work piece, and the cut depth is regulated with a micrometer in the compound rest or carriage.
4. Prior to turning on the machine, cutting parameters such as speed, feed, and depth of cut are selected.
5. Now remove the work piece and use a marking equipment such as a V-block, vernier height gauge, scriber, or surface plate to make the necessary markings on the circular face of the work piece.
6. In order to accomplish an eccentric turning operation, markings are made.
7. The centre punch is used to punch the indicated line.
8. Attach the work item to the 4-jaw chuck and perform a centering operation so that the circular face's centre coincides exactly with the lathe's axis.
9. Eccentric turning operations are performed to obtain the desired shape of the work piece, with each run having a depth of cut of 1mm.
10. The compound rest is swivelled to the angle "theta" in degrees (as computed) with the use of the align key, and the taper turning angle is determined using the taper angle formula. On each run, the taper turning is done at this angle for a length of 30mm and a depth of cut of 0.5mm.

$$\theta = \tan^{-1} \frac{(D-d)}{2l}$$

*Where*

$\theta$ = Swivel the compound rest to the calculated angle.

D = Major diameter.

d = Minor diameter.

l = Length of the taper.

11. Turning the step by varying the depth of cut on each pass, the desired shape of the work piece can be achieved.

## Observations

1. Work piece material: ________________
2. Cutting Tool Material: ________________
3. Feed: ________________ mm/revolution.
4. Depth of Cut ________________ mm.
5. Selected Cutting Speed ________________ m/min.

6. Plain Turning ________________ m/min.
7. Step Turning ________________ m/min.
8. Taper Turning ________________ m/min.
9. Thread Cutting ________________ m/min.
10. Eccentric Turning ________________ m/min.
11. Knurling operation: ________________ m/min.

## Dimensional Accuracy

| | |
|---|---|
| **Diameter in mm** | Given |
| | Obtained |
| **Lengthin mm** | Given |
| | Obtained |

## Result

To perform various operations (turning, thread cutting, knurling, facing operations) in accordance with the provided drawing.

## Precautions

1. Wear a safe outfit.
2. Maintain a safe distance from cutting tools during machining.
3. Work under the supervision of an expert only.
4. Note the readings accurately.
5. Before beginning the experiment, check the machine speed.

## Experiment 3

**Aim:** Making Bolts on a lathe machine.

### Tools Required

- Straight tool holder.
- Centre drill.
- Live centre.
- Stock and die.
- Metal work vice.
- Cutting saw.
- Centre lathe.
- Pedestal grinder.
- HSS tool bit.
- Right-hand tool holder.

### Procedure

1. Allowance for facing is provided when cutting hexagonal material to length.
2. Place one end of the table in the center. The overall length of the lathe is 90 mm.
3. The focal point for the live center drill one end.
4. The hexagonal bar is placed in the three-jaw chuck, with the live centre on the other end. It is held by 5-8 mm.
5. Cut a thread with a diameter of 12 mm and a length of 80 mm – measure size must end at 12 mm minus 0.05 – 0.10 with the micrometer.
6. Fine-tune the tool and tidy up the bolt's head corner.
7. Chamfer the 12-diameter end at a 45-degree angle with a 2 mm radius.
8. Chamfer the head at 45 degrees while holding the 12-diameter in the chuck.
9. Using the Stock and Die, start the thread by holding the 12-diameter. Use cutting compound while using the tailstock to ensure that the axis is square to the die.
10. Remove the thread from the lathe. Finish the thread depth. The thread depth is taken as 60mm here.

### Result

A bolt has been prepared using a lathe machine.

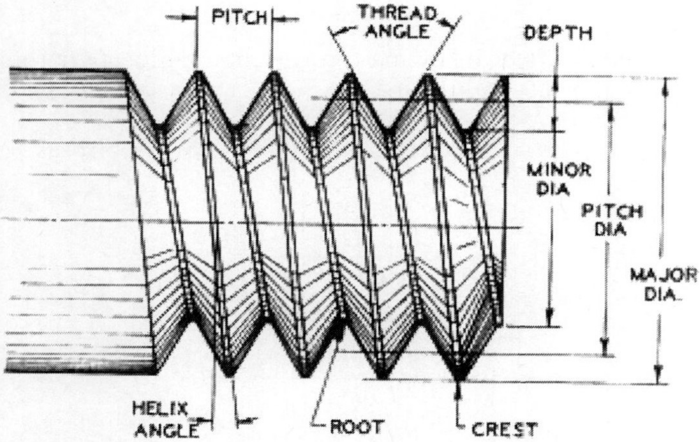


**Figure 270.** Bolt Thread

## Precautions

1. Wear a safe outfit.
2. Maintain a safe distance from cutting tools during machining.
3. Work under the supervision of an expert only.
4. Note the readings accurately.
5. Before beginning the experiment, check the machine speed.

## Experiment 4

**Aim:** To obtain the model illustrated below, perform machine cutting (Knurling, Turning, Facing Operations) on a given metal rod.

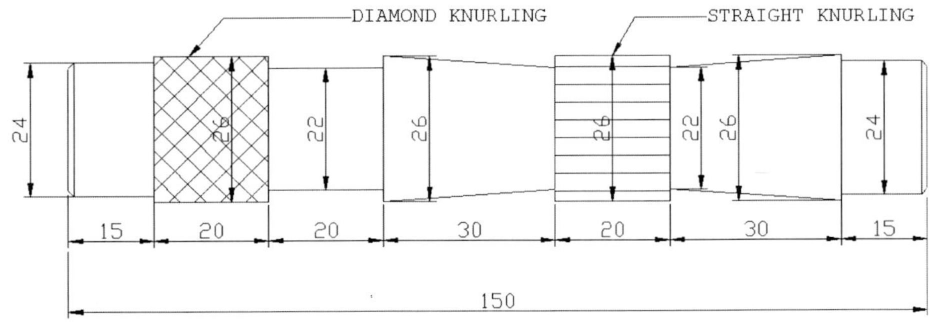


**Figure 271.** Knurling, Turning, Facing Operations

### Tools Required

- Knurling Tools.
- Parting Tool.
- Chuck Key.
- Tool Post Key.
- Side Caliper.
- Right Hand Single Point Cutting Tool.

### Procedure

1. Facing is done on both sides of the piece of art.
2. The work piece is held in the middle, between the live and dead centres.
3. The basic turning operations are completed, and the work piece is marked according to the specifications.
4. The step-by-step turning is completed.
5. Using the appropriate tools, perform taper turning, knurling, and chamfering.
6. The model is now complete.

## Observations

<table>
<tr><td>Specified Diamensions</td><td></td><td></td><td></td><td></td><td></td><td></td><td></td></tr>
<tr><td>Actual Diamensions</td><td></td><td></td><td></td><td></td><td></td><td></td><td></td></tr>
<tr><td>Deffirence</td><td></td><td></td><td></td><td></td><td></td><td></td><td></td></tr>
<tr><td>Allowable Tolerance</td><td></td><td></td><td></td><td></td><td></td><td></td><td></td></tr>
<tr><td>Date of Commencenment</td><td></td><td></td><td></td><td></td><td></td><td></td><td></td></tr>
<tr><td rowspan="2">Date of Completion</td><td rowspan="2" colspan="3"></td><td>Marks</td><td colspan="2">Maximum</td><td>Awarded</td></tr>
<tr><td>Dimensional accuracy</td><td colspan="2"></td><td></td></tr>
</table>

**Figure 272.** Observation Table

## Result

The required model is completed.

## Precautions

1. Wear a safe outfit.
2. Maintain a safe distance from cutting tools during machining.
3. Work under the supervision of an expert only.
4. Note the readings accurately.
5. Before beginning the experiment, check the machine speed.

## Experiment 5

**Aim:** To use a lathe machine to execute thread cutting and knurling on a circular C.S. rod in order to get the design shown in the drawing.

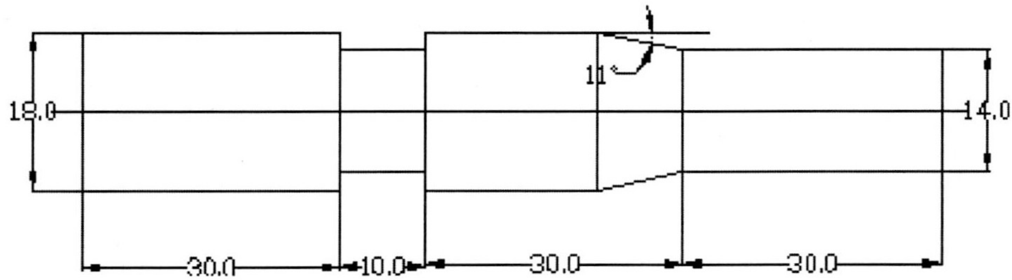


**Figure 273.** Thread Cutting and Knurling on a Circular C.S. Rod

### Tools Required

- Given specimen.
- Thread cutting tool.
- Knurling tool.
- Vernier calipers.
- Pitch gauge.
- Chuck key.
- Tool post key.
- Brush.

### Procedure

1. Use the chuck key to fix the work on the machine. Fix the single-point cutting tool to the task and turn it to the required diameter.
2. Chamfer one end of the edge and undercut the other.
3. Thread the threading process by engaging the bed screw.
4. When the pitch gauge reads zero, come to a halt.
5. Reverse the job and hold it carefully to avoid damaging the threads. The back gear and lead screw must be disengaged.
6. Place the knurling tool against the spinning part of the work.

### Result

The required dimensions of the work piece are obtained.

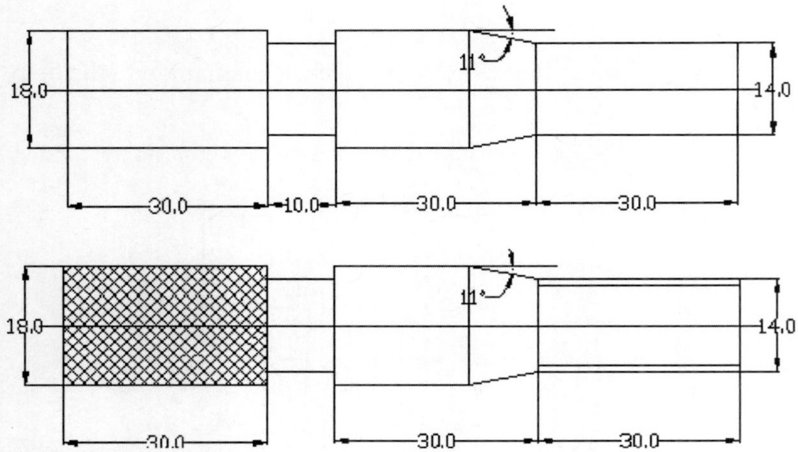


**Figure 274.** Obtained Result

## Precautions

1. Before beginning the experiment, check the machine speed.
2. Before starting the machine, the chuck key must be removed.
3. Feeding should be done in stages.
4. Wear safe outfit.
5. Maintain a safe distance from cutting tools during machining.
6. Work under the supervision of an expert only.

## Experiment 6

**Aim:** To prepare a model according to a given sketch or diagram with the help of lathe machine.

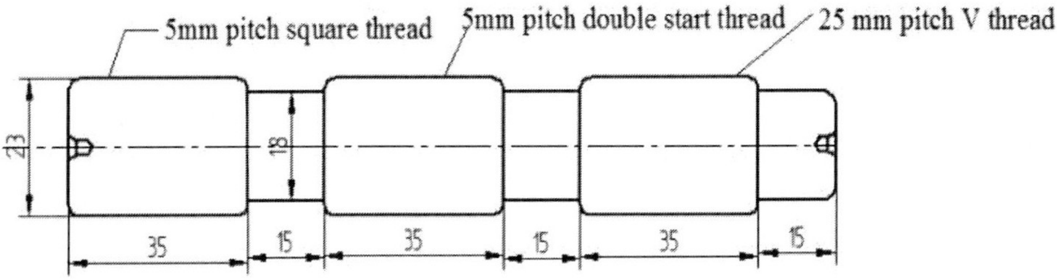


**Figure 275.** Model

### Tools Required

- Single point cutting tool.
- Vernier Caliper.
- Spanner.
- Steel Rul

### Operations needed to be performed

- Turning.
- Taper turning.
- Knurling.
- Facing.
- Countersinking.

### Procedure

1. Check if the given workpiece is suitable according to the dimensions given in the sketch.
2. Drill a counter strike on the face end of the workpiece.
3. Turn the job according to the dimensions given on the sketch.
4. Now the marking should be done as per the sketch.
5. Turn the shoulder as per the sketch.
6. After this, the other operations (Knurling, Taper turning, and threading) should be performed if required.

### Result

The workpiece of desired design/pattern and dimensions is obtained.

**Precautions**

1. Machines should be handled carefully according to the supervision of professionals.
2. Safety equipment like safety goggles, hand gloves and other general instructions should be followed.
3. The machine should be checked before being switched on or used.

## Lathe Machine Objectives Questions

1. It is a small size lathe and fitted at a particular place. It is used for small light-weight work pieces it called __________ .

   a) Speed Lathe  b) **Bench Lathe**

   c) Capstan and Turret Lathe  d) Engine Lathe

2. The foundation of a lathe machine is __________

   a) Headstock  b) Tailstock

   c) Carriage  d) **Bed**

3. A drill chuck is held __________ .

   a) On the Bed  b) **In the spindle of the Headstock**

   c) On the Carriage  d) Both a and c

4. Which of the following is the work-holding device in the lathe machine?

   a) HeadStock  b) TailStock

   c) **Chuck**  d) Bed

5. They are programmable and computerized numerically controlled lathes.

   a) Automatic Lathe  b) Turret Lathe

   c) Capstan Lathe  d) **Special purpose Lathe**

6. It consists primarily of a hollow spindle and a mechanism for driving and controlling spindle speed.

   **a) Headstock**  b) Tool post

   c) Carriage  d) Tailstock

7. It has locations for the tool post or compound rest on the upper face.

   a) Saddle  b) **Cross Slide**

   c) Bed  d) Tailstock

8. The process in which material is removed continuously and accurately with the help of a toothed tool.

   a) Shaping  b) Reaming

   c) **Broaching**  d) Planing

9. The cutting edge stays perpendicular to the tool travel direction in _____ .

   a) Oblique Cutting  b) **Orthogonal Cutting**

   c) Perpendicular Cutting  d) Longitudinal Cutting

10. __________ is a water base cutting fluid.

a) Straight Oils  b) **Soluble Oils**

c) Solid Lubricants  d) Compounded Oils

## Lathe Machine Subjective Questions

1. What is the primary purpose of a lathe? List the several types of lathes.
2. Describe the operation that can be done on a lathe.
3. What is chuck. List various types of chucks used in lathes.
4. What is mandrel ? Why it is used in lathe machine.
5. Differentiate between the motion of the cross slide and the compound rest?

## Lathe Machine Solved Numericals

1. A lathe has four steps the diameter of each been 80 mm, 120 mm, 160 mm and 200 mm. The countershaft pulley revolves at 100 r.p.m. The gears A, B, C and D have 16,48,16,48 teeth respectively. Find the various speeds of the spindle.

**Solution:** Speed without back gear:

As we know that, $n_2/n_1=D_1/D_2$ or $n_2=D_1/D_2 \times n_1$

*Where*

$n_1$ = r.p.m. of the countershaft,

$n_2$ = r.p.m. of the spindle,

$D_1$ = diameter of the countershaft pulley, and

$D_2$ = diameter of the spindle pulley.

**Now,**

a) $n_2 = 200/80 \times 100 = 250$ r.p.m.

b) $n_2 = 160/120 \times 100 = 133.33$ r.p.m.

c) $n_2 = 120/160 \times 100 = 75$ r.p.m.

d) $n_2 = 80/200 \times 100 = 40$ r.p.m.

Speed with back gears: $n_D = n_A \times Z_A/Z_B \times Z_C/Z_D$

**So,**

a) $n_D = 250 \times 16/48 \times 16/48 = 27.8$ r.p.m.

b) $n_D = 133.33 \times 16/48 \times 16/48 = 14.8$ r.p.m.

c) $n_D = 75 \times 16/48 \times 16/48 = 8.33$ r.p.m.

d) $n_D = 40 \times 16/48 \times 16/48 = 4.44$ r.p.m.

**Ans.** Speeds of the spindle are:4.44, 8.33, 14.8, 27.8, 40, 75, 133.33, 250

2. In taper turning operation on a lathe the large and small diameter of taper are are 80 mm and 70 mm respectively and the length of the tapered part is 100 mm. Find the value of conicity (K).

   **Solution:**

   Given – D = 80 mm, d = 70 mm and $l$ = 100 mm
   As we know that, Conicity, i.e. K = (D – d)/$l$
   So, K = (80 – 70)/100 = 1/10
   **Ans.** K=1/10

3. A shaft 1200 mm long has a taper of 1:150 for a length of 300 mm. The maximum diameter of the shaft is 75mm. Determine the minimum diameter of the shaft and the amount setover.

   **Solution:** Length of the taper = 300 mm As we know that,
   Therefore,
   or d = 73 mm Now,
   K = (D – d)/$l$ 1/150 = (75 – d)/300
   Setover = L x (D – d)/2$l$ = 1200 x (75 – 73)/(2 x 300) = 4 mm

   **Ans.** Minimum diameter of the shaft and the amount setover are 73 mm and 4 mm respectively.

4. The pitch of a leadscrew is 5 mm, and the pitch of the thread to be cut is 1.25 mm. Find the change gears.

   **Solution:**
   driver teeth / driven teeth = pitch of the work / pitch of the leadscrew or
   driver teeth / driven teeth = 1.25/5 = 1/4 = (1 x 20)/(4 x 20)

   **Ans:** The driver will have 20 T and the driven gear on the leadscrew 80 T.

5. Find the taper per inch and the taper per foot, if the diameters of the taper are 7/8 inch and 3/8 inch; and the length of the taper is 5 ¼ inch.

   **Solution:**

   Given: Diameters of taper are 7/8 inch and 3/8 inch Len gth of the taper = 5 ¼ inch

   **Now,**
   Taper per inch = (7/8 – 3/8)/5 ¼ = 2/21 Answer.
   Taper per foot = (7/8 – 3/8)/(5 ¼ / 12) = 8/7 Answer.

## Lathe Machine Unsolved Numericals

1. A lathe is used to turn a steel rod with a diameter of 50 mm. The cutting speed is set to 180 m/min. Calculate the required spindle speed (RPM) for this operation.

   **Ans:** Spindle Speed (RPM) = 180 m/min / (π × 0.05 m) H" 1145.92 RPM.

2. A lathe is used to perform a facing operation on a cylindrical workpiece. The workpiece has a diameter of 120 mm, and the tool feed rate is set to 0.25 mm/rev. Calculate the time required to complete the facing operation if the length of the cut is 150 mm.

   **Ans:** Time = 150 mm / (0.25 mm/rev × 1) = 600 rev = 600 seconds.

3. A lathe machine is used to turn a workpiece from an initial diameter of 80 mm to a final diameter of 60 mm. If the feed rate is set to 0.2 mm/rev, calculate the total number of revolutions required for turning.

   **Ans:** Total Revolutions = (80 mm - 60 mm) / (2 × 0.2 mm/rev) = 100 revolutions.

4. A lathe machine is used to thread a steel rod with a major diameter of 30 mm and a pitch of 3 mm. Calculate the lead of the thread.

   **Ans:** Lead of the thread = 3 mm.

5. A lathe machine is used to turn a cylindrical workpiece with a diameter of 160 mm. The tool feed rate is set to 0.15 mm/rev, and the spindle speed is 800 RPM. Calculate the cutting speed.

   **Ans:** Cutting Speed = π × 160 mm × 800 RPM H" 4019.69 mm/min.

# 8

# Shaper Machine

## Introduction

Shaping machine, shaping machine is a heavy commercial machine used for generating flat surfaces on the workpiece as per the requirement. It is specifically designed to give the desired shape to the workpiece which can be flat, horizontal, or vertical as per the requirement.

The shaper machine is a reciprocating type of machine (that is, the machine that converts the rotatory motion into the linear motion) in which the tool reciprocates with the forward and backward stroke removing the metal form of metal chips and turning the Metal workpiece into the desired shape.

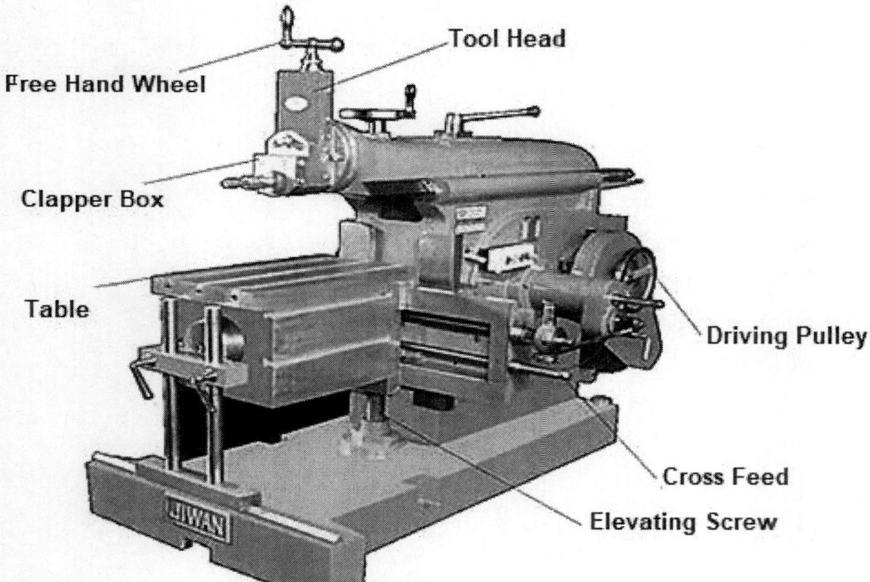


**Figure 276.** Parts of Shaper Machine

## Parts of Shaper Machine

1. **Base:** It joints with concrete, and it is made up of cast iron, has compressive strength, and tolerates high weight.
2. **Column:** It is attached to base, it has all gear arrangements, drive mechanism or we can say that it provides housing to the mechanism.
3. **Guide Ways:** These are attached to columns (one is at the side and another one at the upper side of the column) which provide motion to Ram and the table.
4. **Table Sliding Support:** It provides support for a table to move up or down.
5. **Elevating Handle:** It is used for moving tables.
6. **Vise:** It consists of jaws (one is movable, and another is free) which are used for holding the work piece tightly.
7. **Ram:** It consists of all the cutting tool assembly, and it moves forward and backward to a crank and Lever mechanism.
8. **Ram Clamp:** Sets the position of the ram.
9. **Tool Head or Vertical slide:** It gives the motion to clapper the box.
10. **Tool Feed Handle**: It is used to move the tool head.
11. **Clapper Box:** It gives support to fix the tools with the help of a tool post.
12. **Tool Post:** Helps to hold tools.
13. **Cross feed Handle:** It helps to move table cross wards.

## Specification

1. Smash cycle every second or stroke every second 15 to 90.
2. Vertical journey of honk slide one hundred fifty mm.
3. Degree journey of desk 450 mm.
4. Size of facet desk 473mm x 330 mm.
5. Ground area essential 1981 mm x 1067mm.
6. Vertical journey of desk 475m.
7. Internet weight 1750 kg.
8. The pressure of engine 3HP.

## Construction

The Shaper machine's key components are the Base, Body (Pillar, Frame, and Column), Cross rail, Ram, and tool head (Tool Post, Tool Slide, Clamper Box Block). The foundation is manufactured from hefty cast iron and is secured to the shop floor. It facilitates the system's normal load similarly to the body structure. Vibrations and exceptional forces which can be likely to be created are absorbed and withstood through the lowest.

**Body**: It is hooked up on the lowest and houses the strength component packing the primary drives, the stuff subject and the short move returned tool for the damage development. The problem of the body gives guide techniques to the damage and it's the front gives the side techniques to the passing rail.

**Cross Rail**: its miles hooked up on the facade of the body outline and can be lengthy beyond proper right here and there. The upward development of the passing rail gives occupations of several statures to be obliged beneath the apparatus. Sliding along the passing rail is a seat that conveys the artwork desk.

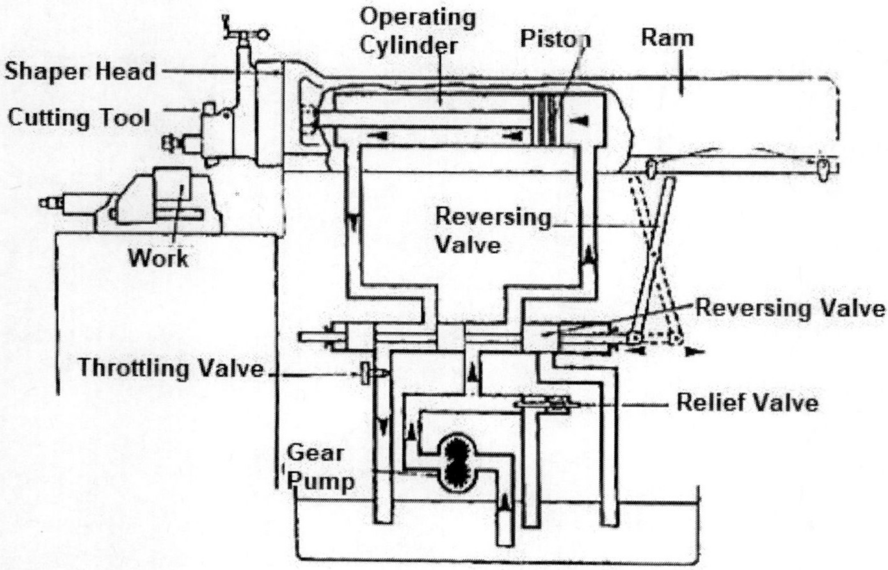


**Figure 277.** Construction Details of Shaper Machine

**Ram and Tool Head**: The ram is driven to and fro in its slides through the opened connection tool. The to and fro movement developed is known as a stroke and it has an inclination to be changed through the length of the work piece to be machined.

## Working

There is a single factor slicing device utilized in this system and it has the most effective slicing edge. The turning device is a great instance of a single factor slicing device. It is positioned inside the device holder and additionally installed at the ram. The work piece is clamped without delay at the desk for machining and it could assist at any other end. Ram is showing the reciprocating movement. And the slicing device holder is moved from side to side on the floor of the work piece. Many of you believe you studied that the ram rotates and slices in strokes. And it's far from relying on the shapes. The slicing takes region in the forwarding stroke in a popular shaper system and the backward stroke is taken into consideration to be idle. The slicing devices were given movement from the fast go back mechanism and slicing intensity is accomplished through a motion of the device. Feed movement follows the pawl and ratchet mechanism.

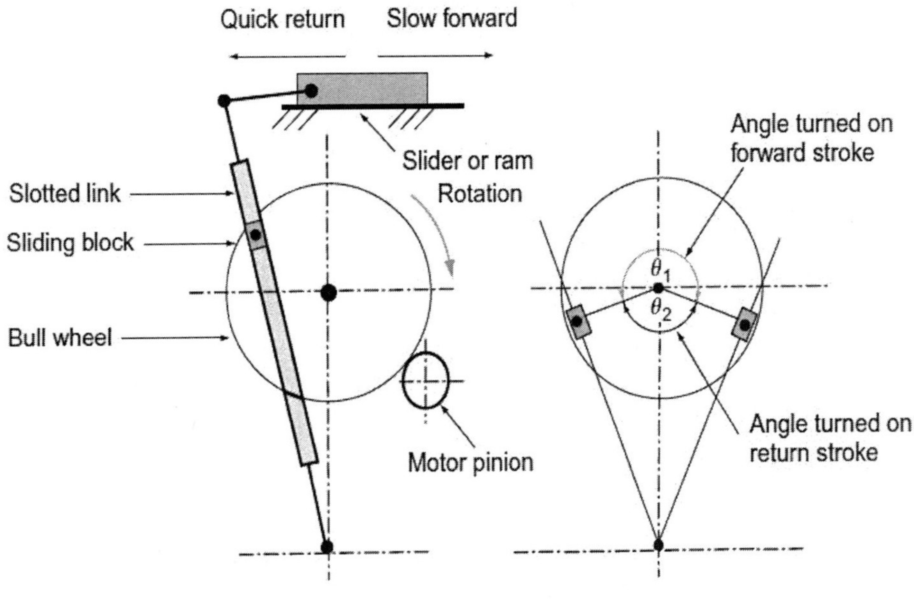


**Figure 278.** Crank and Shaft Mechanism

## Types of Shaper Machines

### 1. Based on the type of driving mechanism

i) **Crank Type Shaper**: Crank and slotted link mechanism of a crank type shaper converts the rotation of an electric motor into reciprocating movement of the ram. Though the lengths of both the forward and return strokes are equal, the ram travels at a faster speed during return stroke. This quick return is incorporated in almost all types of shaper.

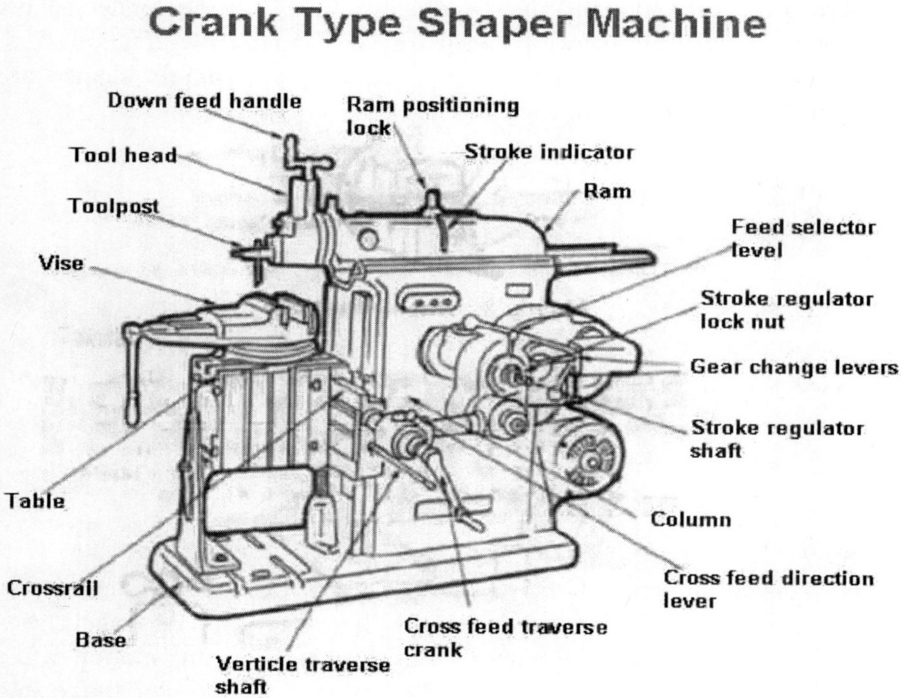


**Figure 279.** Crank Type Shaper

## ii) Geared Type Shaper:

**Figure 280.** Geared Type Shaper

iii) **Hydraulic Type Shaper**: The ram of a hydraulic shaper is connected to a piston. Oil at high pressure is pumped to the cylinder of the hydraulic system. As the oil pushes the piston, the ram reciprocates. Hydraulic shapers are high power machines and are used for heavy duty work.

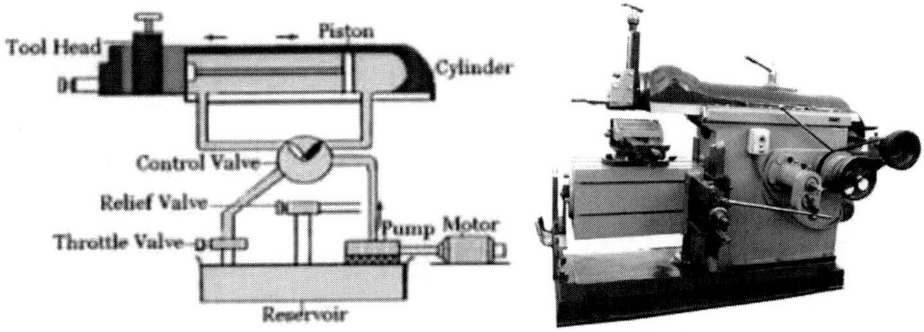


**Figure 281.** Hydraulic Type Shaper

## 2. Based on Ram Travel

i) **Horizontal Shaper**: A horizontal shaper is a type of machine tool used in metal working and machining processes. It's designed to cut flat surfaces on a work piece, typically metal, by using a reciprocating motion of the cutting tool. The work piece is usually clamped to a table, and the cutting tool moves back and forth horizontally to shape the material. The horizontal shaper is capable of producing various shapes, angles, and contours on the work piece.

ii) **Vertical Shaper**: A vertical shaper is another type of machine tool used for shaping metal or other materials. In this case, the cutting tool moves in a vertical direction, meaning it moves up and down to create cuts on the work piece. The work piece is generally mounted on a table that can move horizontally to position it correctly under the cutting tool. Vertical shapers are often used for shaping internal surfaces, keyways, and slots.

**Figure 282.** Horizontal Shaper

**Figure 283.** Vertical Shaper

### 3. Based on the Table Design

i) **Universal Shaper:** A universal shaper is a type of shaper machine that is designed with the capability to perform a wide range of shaping operations. Unlike specific-purpose shapers that are designed for a particular type of work, a universal shaper is versatile and can be adapted to various tasks. It often comes with additional features and attachments that allow it to perform not only horizontal shaping but also vertical shaping, angular shaping, helical or spiral shaping, and even more intricate operations. The flexibility of a universal shaper makes it suitable for producing complex shapes, contours, and profiles that might be challenging to achieve with a standard shaper. However, these machines tend to be more complex and may require skilled operators to set up and operate effectively due to the various adjustments and configurations they offer.

ii) **Standard Shaper:** A standard shaper, often simply referred to as a "shaper," is a basic machine tool designed primarily for linear, horizontal shaping operations. It consists of a reciprocating cutting tool that moves horizontally across the work piece, removing material and producing flat surfaces, keyways, slots, and other linear features. Standard shapers are more straightforward in design compared to universal shapers and are commonly used for shaping tasks that don't require intricate or complex operations. Standard shapers are reliable and easier to operate than universal shapers, making them suitable for many general machining applications. However, they might lack the versatility needed for tasks that involve cutting at different angles, producing helical shapes, or creating more elaborate profiles.

**Figure 284.** Universal Shaper

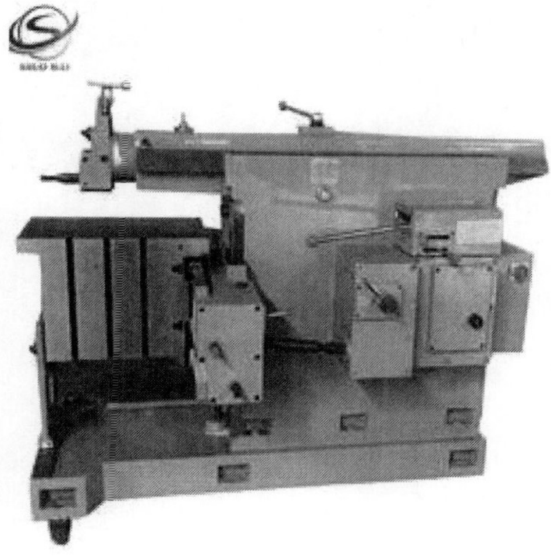

**Figure 285.** Standard Shaper

## 4. Based on the Cutting Stroke

i) **Push Cut Type Shaper:** In a push cut type shaper, the cutting tool is pushed away from the operator (or the workpiece is pulled toward the tool) during the cutting stroke. This means that the cutting tool moves against the direction of the workpiece feed. The cutting action takes place as the tool engages with the workpiece, removing material as it moves. Push cut type shapers are often used for lighter cuts and finer surface finishes. The tool is typically set to take a shallow cut during the forward stroke.

ii) **Draw Cut Type Shaper:** In a draw cut type shaper, the cutting tool is pulled towards the operator (or the workpiece is pushed away from the tool) during the cutting stroke. This means that the cutting tool moves in the same direction as the workpiece feed. In this type of cut, the tool engages with the workpiece more aggressively, removing larger amounts of material during each pass. Draw cut type shapers are generally used for heavier cuts and roughing operations. The tool is set to take a deeper cut during the forward stroke.

**Figure 286.** Push Cut Type

**Figure 287.** Draw Cut Type

## Size of Shaper Machine

1. The size of a shaper is decided by the maximum length of stroke or cut it can make.
2. The standard size goes from **175 to 900 mm.**
3. The length of stroke shows, in addition to the general size of the machine, the size of a **3D** square that can be held and planed in the shaper.
4. Consequently in a **250 mm** shaper the length of stroke might be changed from **0 to 250 mm.**
5. The cross feed adjustment of the table will be **250 mm** and the extreme bottom position of the cross rail will permit the table to accommodate a work piece **250 mm** high.
6. The length of stroke of a shaper simply shows the general size of the shaper.
7. Different points of interest, such as the type of drive, belt drive, or individual motor drive, power input, floor space required, weight of the machine, cutting to return stroke ratio, number and amount of feed, etc. are also sometimes necessary.

## Pros of Shaper Machine

1. In the case of a shaper machine, a single-edged cutting tool is hired which results in low machining cost.
2. The set-up is exceptionally speedy and simple and furthermore can be promptly changed starting with one occupation then onto the next work.

## Cons of Shaper Machine

1. As per the mechanism, the shaper machine can be considered a lethargic machining tool because of its reciprocating movement limitation due to which the output requires a greater number of operational strokes.
2. The cutting velocity isn't typically extremely high rates of responding movement because of the high idleness power created in the movement of the units and parts of the machine.

## Working Mechanism

The ram moves at a comparatively slower speed during the forward cutting stroke. During the return stroke, the mechanism is so designed to make the tool move at a faster rate to reduce the idle return time. This mechanism is known as quick return mechanism.

At the ram moves at a faster rate during stroke, the time taken becomes less. The total machining time decreases and the rate of production increases. The following mechanisms are used for quick return of the ram.

1. Crank and slotted link mechanism
2. Hydraulic mechanism
3. Whitworth mechanism

## 1. Crank and Slotted Link Mechanism

An electrical motor runs the driving pinion(S) at a uniform speed. This pinion makes the bull gear (M) to rotate at a uniform speed. Bull gear is a large gear fitted inside the column. The point 'O' is the centre of the bull gear. A slotted link having a long slot along its length is pivoted about the point 'K'. A sliding block 'N' is fitted inside the slot and slides along the length of the slotted link. 'P' is the crank pin and 'OP' can be considered as a crank. Figure 285 shows the crank & slotted link mechanism.

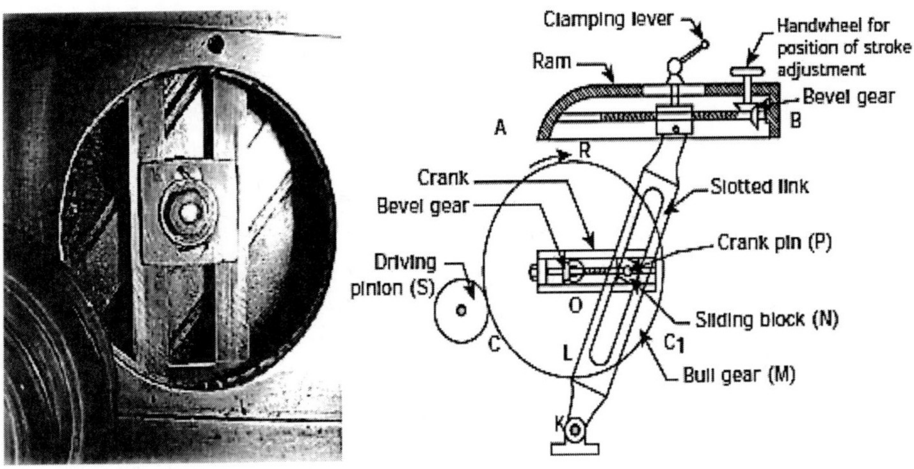


**Figure 288.** Crank and Slotted Link Mechanism

When the bull gear rotates, the sliding block also rotates in the crank pin circle. This arrangement provides a rocking movement to the rocker arm. As the top of the slotted link is connected to the ram, the ram reciprocates horizontally. So, bull gear rotation is converted into the reciprocating movement of the ram.

## 2. Quick Return Mechanism

As shown in the diagram, 'KA' indicates the starting point of the forward cutting stroke and 'KB' the end of the cutting stroke. The rotation of the crank 'OP' in clockwise direction through the angle CRC1 refers to the forward cutting stroke.

The rotation of the crank in the same direction through the angle C1LC refers to the return stroke. As the angle C1LC is smaller than the angle CRC1, the time taken for the return stroke is less than that of forward stroke. So, it is evident that the speed at which the ram travels during return stroke is more.

$$\frac{\text{Time taken for forward cutting stroke}}{\text{Time taken for the idle return stroke}} = \frac{\text{angle } CRC_1}{\text{angle } C_1LC}$$

$$= 216^{\circ} / 144^{\circ}$$

$$= 3 / 2,$$

**In some machines this ratio can be set as 7/5.**

**Figure 289.** Observation

The stroke length of a ram is the distance the ram moves forward or backward. It depends upon the distance between the centre of the bull gear and the centre of the sliding block. It is adjusted according to the length of the work.

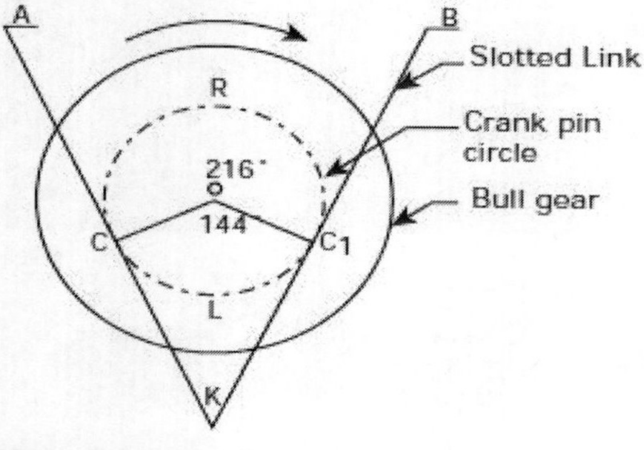


**Figure 290.** Quick Return Mechanism

## Work Holding Devices

Work pieces can be held and supported on the shaper table directly or by having some special devices. Depending on the size and shape of the work, it may be supported on the table by any one of the following methods.

1. Shaper vise
2. Clamps and stop pins
3. T-bolts and step blocks
4. Angle Plate
5. V-Block
6. Special fixtures

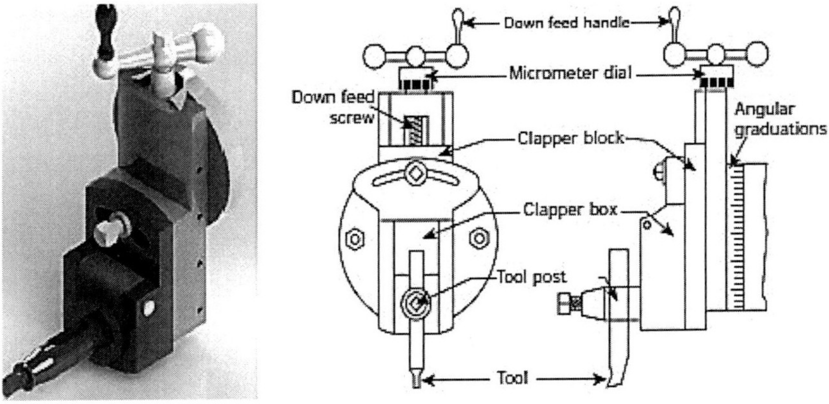


**Figure 291.** Swivel Tool Head

## Vise

Vise is the most common and simple work holding device used in a shaper. Different types of vises are used in a shaping machine according to the need and they are: 1. Plain Vise, 2. Swivel Vise, 3. Universal Vise.

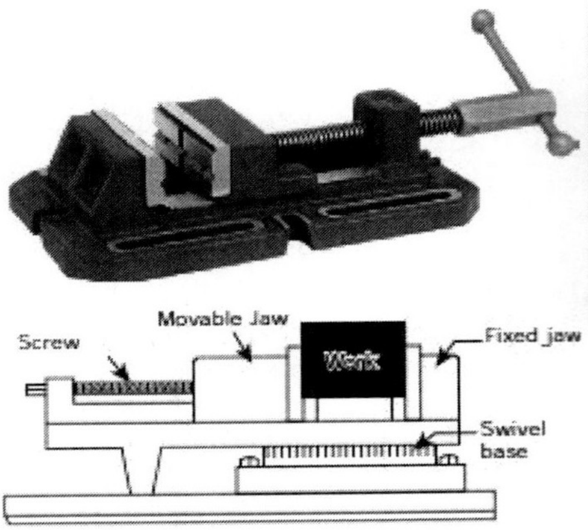


**Figure 292.** Vise

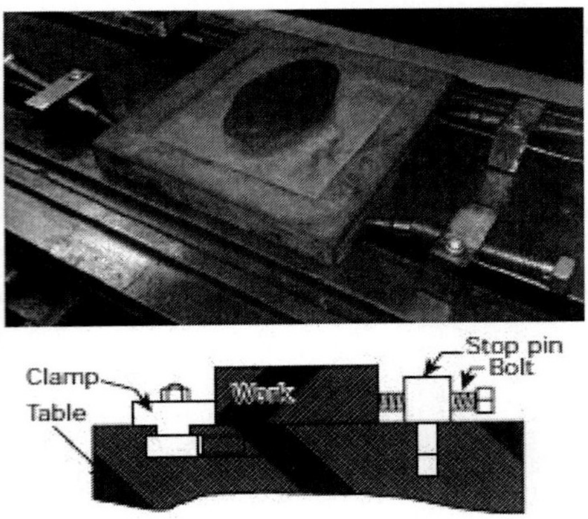


**Figure 293.** Clamp and Stop Pin

### Clamps and Stop Pins

T-bolts are fitted into the T-slots of the table. The work is placed on the table. The work is supported by a rectangular strip at one end and by a stop pin at the other side. The screw is tightened to secure the work properly on the machine table. The use of stop pin is shown in Figure 293.

### T-Bolts and Step Blocks

The step blocks are used in combination with T-bolts and clamps to hold the work directly on the machine table. The t-bolts are fitted in the T-slots of the machine table. One side of the clamp holds the work and the other side rests on a step of the step block. The different steps of the block are useful in leveling the clamp when holding works of different heights. A nut on the top of the clamp holds the work rigidly. 'T' bolt, clamp & step block.

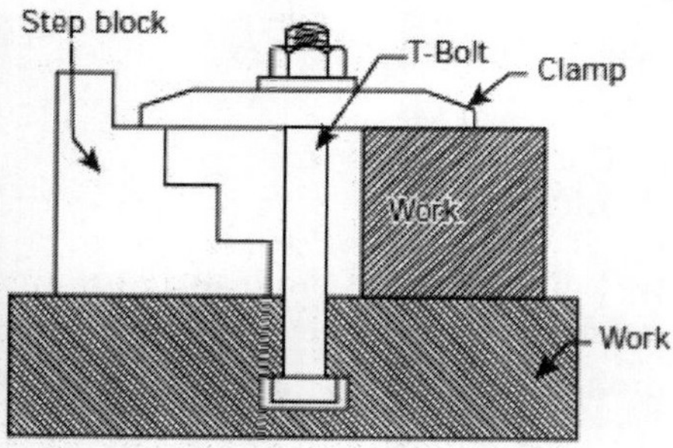


**Figure 294.** T Bolt and Clamp

### Angle Plate

Angle plate resembles the English alphabet 'L'. It is accurately machined to have two sides at right angles. Slots are provided on both sides. One of the sides is bolted to the machine table and the work piece are held on the other side. The use of an angle plate is shown in Figure 295.

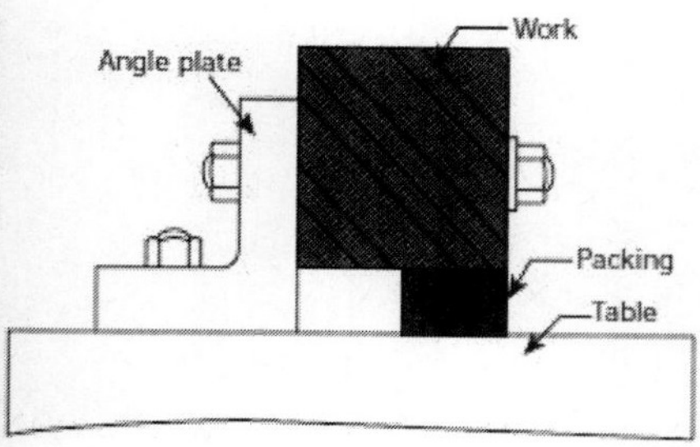


**Figure 295.** Angle Plate

## V-Block

V-block is a metal block having a 'v-shaped groove on it. It is used for holding cylindrical workpieces. Operations like keyway cutting, slot cutting, and machining flat surfaces can be performed on the cylindrical work pieces held on a 'v' block. The use of a 'V' block is illustrated in Figure 296.

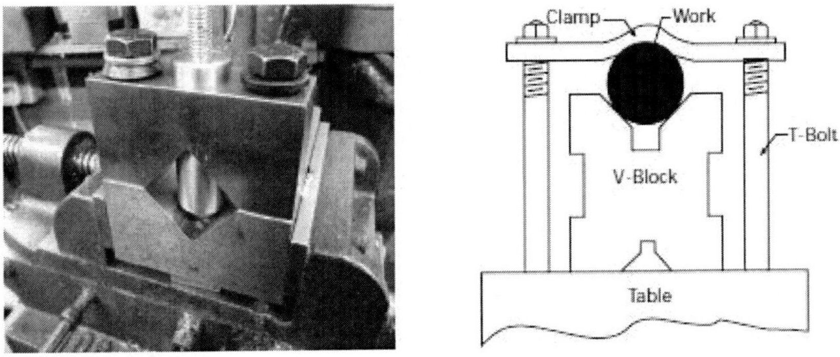


**Figure 296.** V Block

## Special Fixtures

When internal keyways are to be machined on the holes, the work is held with a special fixture. The fixture has a V-block attached to it and the cylindrical work is mounted on it.

## Experiments 1

**Aim:** To study the Horizontal Surface Operation of a shaper machine.

**Materials Required:** A shaper machine with a single-point cutting tool.

**Theory:** It's one of the most common procedures on a shaper machine. The work is fed horizontally under the reciprocating tool in this method, and the surface generated is horizontal and flat.

### Procedure

1. Depending on its size, the work is fastened directly on the machine table or held in a vice. The tool is securely housed in a tool holder. It is positioned at a suitable slant and height above the task.
2. Until the cut begins, cross feed to the table is done by hand. Following that, a power feed can be used. The machine is halted after the cut is completed, and the work is inspected. The operation is repeated if more material has to be removed until the desired surface is achieved.
3. Setting the instrument for horizontal cutting necessitates extra consideration.
4. Another thing to keep in mind when buying a toolset is that the cutting edge should not protrude too far below the tool holder. The tool head's vertical slide should not be allowed to overhang too much below the ram. The tool will be weakened and put under strain if this does not happen. There will also be a lot of buzzes.

   **Result:** Horizontal surface is obtained.

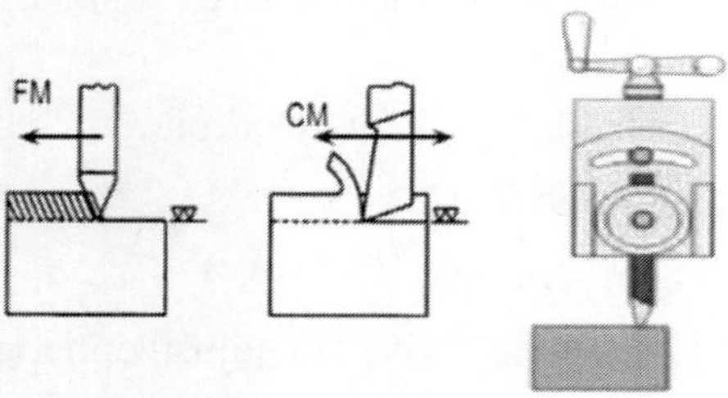


**Figure 297.** Shaper Machine

### Precautions

1. Work under the supervision of an expert
2. Maintain a safe distance from a machine during the operation
3. Work with the proper safety equipment.
4. Wear all safety things and dress
5. Not move anywhere leaving the machine running.
6. Go through all the instructions and process steps once again properly.

## Experiments 2

**Aim:** Machining a Block on a shaper machine.

**Materials Required:** Mild steel blocks will be used which have sides greater than 75 mm on each side.

## Tools Required

- Shaper machine.
- Vernier caliper.
- Try square.
- Single point cutting tool.
- Vise.

**Theory:** As this operation is done using mild steel blocks, it is a general operation performed by a shaper machine.

## Procedure

1. Required dimensions are given and the job was checked more than once to the dimensions.
2. One side of the block is marked 75 mm and punching was done and then every side got punched according to the Mark.
3. The job was properly attached in the vise of a shaper machine.
4. Mild steel and shaper were set according to the calculation taken in stoke / mm.
5. Then the stroke length was fixed as per the length of the workpiece.
6. In a tool holder, a single cutting point tool was fixed.
7. Now initiate the machine for the cutting or shaping.
8. After finishing one side, the surface is upturned and again adjusted in the vise.
9. Then the job is checked for perpendicular and parallel positioning.
10. The job was checked for dimensions.
11. Then the square of the 75 mm side was obtained in the shaper machine from the stock.
12. Then after removing the workpiece and cleaning the burned surfaces the accuracy of the required shape of the workpiece was checked.

    **Result:** Thus the desired shape of the block obtained.

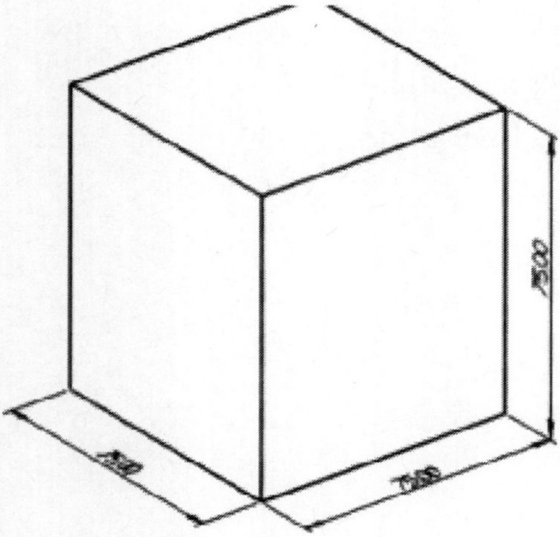

**Figure 298.** Drawing of Block

## Precautions

1. Work under the supervision of an expert.
2. Maintain a safe distance from a machine during the operation.
3. Work with the proper safety equipment.
4. Wear all safety things and dress.
5. Not move anywhere leaving the machine running.
6. Go through all the instructions, processes, and steps once again properly.

## Experiments 3

**Aim:** Making V-shape on a block in shaper machine.

**Materials Required:** Mild steel blocks having sides 70 x 30 x 75 mm sides.

### Tools Required

- Shaper machine.
- Vernier caliper.
- Try square.
- Single point cutting tool.
- Vise.

**Theory:** We can make V- shape in blocks by using this machine. This machine helps us to make different shapes very easily.

### Procedure

1. Mark a layout line on a piece with holding job in a horizontal line by this layout line and then machining the part by regular feed.
2. Through supporting work on tapered parallel in a vice.
3. When the power hack saw, cut the workpiece in the required dimensions.
4. Then the job is adjusted as per requirement in a vice of the shaper machine.
5. Then, the top surface is machined first.
6. The surface of the job is upturned and clamped firmly in the vice.
7. Then all other remaining surfaces are machined in order.
8. Then the marking is done for making grooves but as per the drawing.
9. By using a round nose cutting tool the groves are formed by machining.
10. Then the grooved feed is produced on the table simultaneously by the given tool.
11. For achieving the desired depth, the grooves are shaped in different type of cuts.

**Result:** Thus the V-Shape on the rectangular block is obtained.

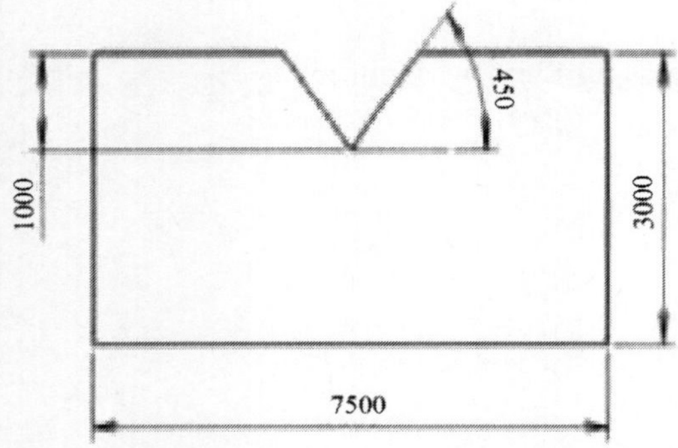


**Figure 299.** V-Shape

## Precautions

1. Work under the supervision of an expert.
2. Maintain a safe distance from a machine during the operation.
3. Work with the proper safety equipment.
4. Wear all safety things and dress.
5. Not move anywhere leaving the machine running.
6. Go through all the instructions, processes, and steps once again properly.

## Experiments 4

**Aim:** To study the Keyway cutting using Shaper Machine.

### Materials Required

- Metal Workpiece.

### Tools Required

- Shaper machine.
- Vernier caliper.
- Try square.
- Single point cutting tool.

**Theory:** This operation is generally done for various purposes and this machine helps to make internal keyways cutting of a block.

### Procedure

1. The workpiece is held in a vice by using a V- block.
2. Then, with the help of a dial indicator, the movement of the stroke is parallel to the axis of the shaft.
3. Hence the shaft position is adjusted and done by a dial indicator.
4. Check if the workpiece is properly held.
5. Then the keyway cutting process is carried out.
6. This process is similarly carried out to the cutting groove process on a shaper.

   **Result:** Thus the keyway cutting of a block is achieved.

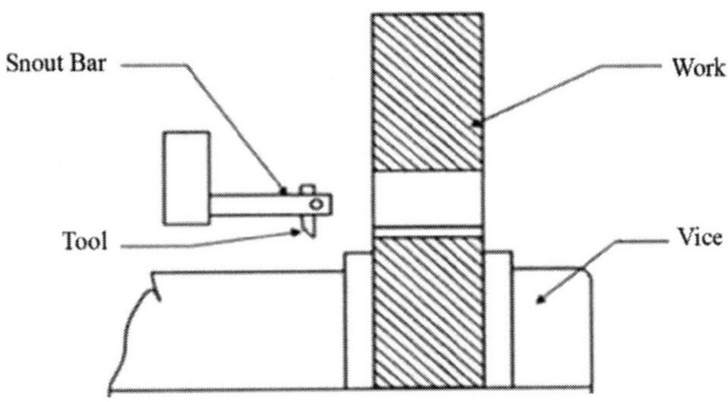


**Figure 300.** Machining of Internal Keyway

## Precautions

1. Work under the supervision of an expert.
2. Maintain a safe distance from a machine during the operation.
3. Work with the proper safety equipment.
4. Wear all safety things and dress.
5. Not move anywhere leaving the machine running.
6. Go through all the instructions, processes, and steps once again properly.

## Experiments 5

**Aim:** Practice of changing a Round MS Rod into a Square Section on a shaper.

### Tools Required

- Metal Rod
- Shaper machine.
- Shaper tool.

**Theory:** This machine works efficiently and helps us to make different types of shapes. Similarly in this operation, it helps us to change a round MS rod into a squares section.

### Procedure

1. Take a horizontal shaper machine.
2. Fix the workpiece on the table with the help of vice.
3. Set the forward and backward strokes of the ram.
4. Select the appropriate shaper tool and fix the tool into the tool head.
5. The tool should be perpendicular to the workpiece.
6. The suitable depth of the cut is taken such that 1.5 mm.
7. Check the squareness of the workpiece. If the operation is not complete then switch on again the shaper.
8. Hence if the required squareness is reached then switch off the shaper.

   **Result:** We get the required Square Section.

### Precautions

1. Work under the supervision of an expert
2. Maintain a safe distance from a machine during the operation
3. Work with the proper safety equipment.
4. Wear all safety things and dress
5. Not move anywhere leaving the machine running.
6. Go through all the instructions, processes, and steps once again properly.

## Experiments 6

**Aim:** To study and carry out the various operations on the shaper machine.

**Theory:** Surfaces designed by shaper machine are as follows:

- Horizontal plain surface.
- Vertical plain surface.
- Inclined surface.
- Grooved surface.
- Slotted surface.
- Stepped surface

## Procedure

1. Check the proper functioning and alignment of the machine.
2. Adjust the height of the table and the saddle.
3. Mount the workpiece on the vice and lock the workpiece through the vice lever.
4. Adjust the height of the tool post and the cutting tool.
5. Adjust the cutting tool according to the operation to be carried out and adjust its inclination as per the requirement.
6. Check the speed gear of the machine.
7. Start the operation of the machine by switching the electric motor.

   **Result:** Desired surfaces are obtained.

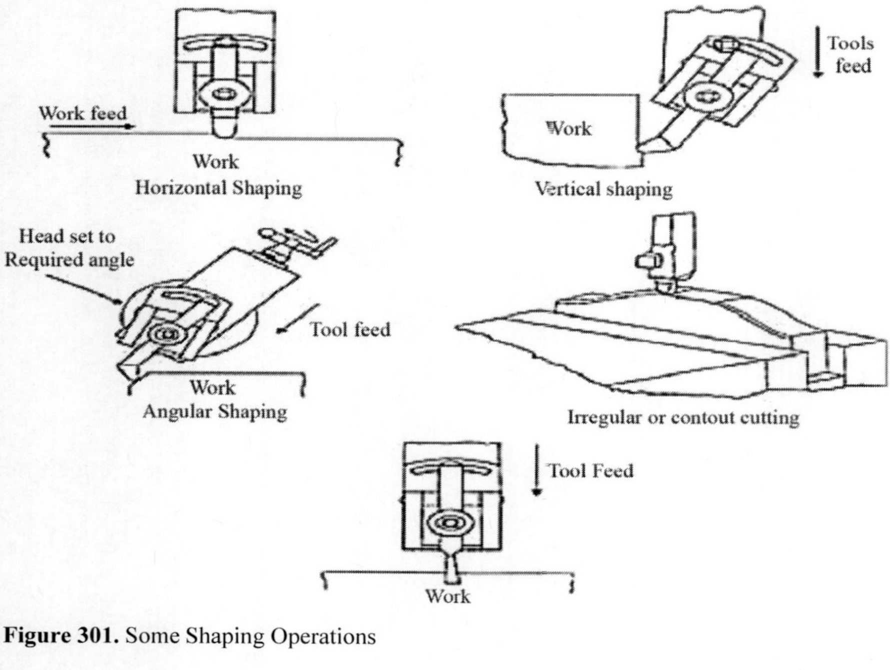


**Figure 301.** Some Shaping Operations

## Some Precautions

1. Work under the supervision of an expert.
2. Maintain a safe distance from a machine during the operation.
3. Work with the proper safety equipment.
4. Wear all safety things and dress.

## Shaper Machine Objective Questions

1. The shaper is a reciprocating type of machine tool primarily intended for producing which surface.
   a) **Flat Surface** b) Angular Surface
   c) Irregular Surface d) None
2. Shaper is termed as standard when the table has ________ movements.
   a) Only One b) **Only Two**
   c) Only Three d) None of the mentioned
3. A quick return mechanism is used in a ________ .
   a) Milling Machine b) Broaching Machine
   c) Grinding Machine d) **Shaping Machine**
4. Hydraulic type shaper is the type of shaper according to ________ .
   a) **Design of the Table** b) Position and Travel of the Ram
   c) Type of Cutting Stroke d) None of the mentioned
5. Cutting of material during slotting operation takes place in ________ .
   **a) Forward Stroke**
   b) Backward Stroke
   c) Both Forward and Backward Strokes
   d) None of the mentioned
6. Traveling head type shaper is a type of shaper according to ________ .
   a) Design of the Table b) **Position and Travel of the Ram**
   c) Type of Cutting Stroke d) None of the mentioned
7. The change of the movement of the drive is done by the mechanism contained within the ________ of the machine.
   a) Cross Rail b) **Column**
   c) Tool Head d) None of the mentioned
8. To reduce the total cutting time, the standard shaper is designed to complete forward stroke ________ .
   **a) Slowly** b) Faster
   c) At Moderate Speed d) None of the mentioned
9. In the crank and slotted link mechanism, an electric motor————.
   **a) Rotates at a Constant speed** b) Rotates at Variable speed
   c) Doesn't Rotates d) None of the mentioned

10. The base of the standard shaper is able to ________ .

    a) Resist Vibration

    b) Take up High Compressive

    **c) Both Resist Vibration and take up High Compressive**

    d) None of the mentioned

## Shaper Machine Subjective Questions

1. With a neat sketch show various components of a shaper and explain their functions.
2. Explain the working and construction of the tool head.
3. Explain the quick return mechanism through sketch.
4. Explain the automatic table feed mechanism for a shaper.

## Shaper Machine Solved Numericals

Q1 Calculate the Power Required For Shaping Stect With A Depth Of Cut Of 2.8mm, Cutting Speed 65 m/Min and The Work Length 50mm. The Feed Rate is 0.5mm/Rev. Take machining constant K As 70x10—6.

**Solution:** Given Data:

T = 2.8mm, V = 65m/Min = 65x1000mm/Min L =50mm, F = 0.5mm/Rev

Material Removal Rate, W = F T V = 0.5x2.8 X
65 X 1000 P = Kw

Power Required, P = 79 X 10-6+ X 91000

= 70189 H.P

= 7.189 X 0.736

= 5.29 Kw

**Ans.** Power required is 5.29Kw.

Q2 Find the machining time required for machining the surface 600 × 800 mm, on a shaping machine. Assume the cutting speed as 8 m/min. The return to cutting time ratio is 1:4, and the feed is 2 mm/double stroke. The clearance at each end is 70 mm.

**Solution:** Lj = 600 mm, w = 800 mm, v = 8 m/min, m = ¼, f = 2 mm/stroke, c = 70 mm Given clearance at each end: c=75 mm

Length of stroke: L = 600 + 2×70 = 740 mm Assume 8.65 strokes/min available

Machining time for 800 mm width t = w/fN = 46.25min Answer: Machining time is 46.25 min.

Q3 Estimate the shortest machining time required in a shaper to machine a plate of 200 * 90mm. under the following conditions.

Cutting Speed = 13.3m/Min Feed = 0.57 Mm/Double Stroke Number Of Passes = One Approach + Overrun (Longitudinal) =20mm (Lateral) = 4mm Ratio of Cutting Speed To Rapid Return = 0.83

**Solution:** Given Data:

Cutting Speed, V = 13.3m/Min

Feed, F = 0.5mm/Double Stroke

Number Of Pass, N = 1

Approach Over Run, A = 20mm

Work Length, L = 200mm

Work Width W = 90mm

Ratio M = 0.83

Cutting speed V= Length of cutting stroke/ Time taken for the same stroke

Total double stroke time= Time for forward stroke + return stroke

T=L/V +m(L/V) =L/V(1+m)

Total length L= l+A =200+20 =220mm = 0.22m

Total time T = 0.22/13.3 (1+0.83) =0.0303min

Number of strokes SN = w/f

Total Width=W+ Lateral Approach = 90+4 = 94mm

SN= 94/0.57 =164.9 = 165 stroke

Total time required for finishing the complete job

Tt=T*SN= 0.0303* 165 = 4.9975mm= 5min

**Ans:** Shortest machining time is 5 min.

Q4 The cross-feed on a shaper consists of a lead screw having 0.2 threads per mm. A ratchet and pawl on the end of the lead screw are driven from the shaper crank such that the pawl indexes the ratchet by one tooth during each return stroke of the ram. Ratchet has 20 teeth. a) find the crossfeed in mm. b) if a plate 100 mm wide has to be machined in 10 minutes find the cutting speed in m/s. The ratio of return to cutting speed is 2:1 and the length of the stroke is 150mm.

**Solution:** Given Data:

| | | |
|---|---|---|
| Lead Screw Threads | = | 0.2 Thread/mm |
| Ratchet Teeth | = | 20 |
| Plate Wide, W | = | 100 mm |
| Machining Time, T | = | 10min |
| Ratio of Return To Cutting Speed 1/M | = | 2 |
| Stroke Length L | = | 150mm |
| Pitch of the Lead Screw | = | 1/lead screw threads |
| | = | 1/0.2 = 5mm |

Pawl Indexes 1/20 revolution per each stroke

Cross Feed: F = Pawl Indexing × Lead Screw Pitch =5/20 =0.25mm

Number Of Cutting Strokes, SN = w/f= 100/0.25 =400

Total Time Required To Complete The Job,

Tt=T* SN =10*400= 400min

Cutting speed, V= Length of cutting stroke/Time taken for the cutting stroke

Machining time T=L/Nf

4000=100/N*0.25

N= 0.1rpm

Cutting speed, V=LN (l+m)/1000 =150*0.1(1+0.5)/1000 = 0.02225m/min.

**Ans.** Crossfeed = 0.25mm and Cutting speed = 0.02225 m/min.

Q5 A 75cm* 20 cm surface of cast iron block of 15 cm thick is to be machined on a shaper. Ram speed is 25m/min, length of stroke is 500 mm, depth of cut is 4 mm, feed is 1.5mm/stroke, stock to be removed is 7 mm side cutting edge angle of the tool is 450, a ratio of time taken in return stroke to the time taken in cutting stroke is 0.5. Determine the metal removal rate and the cutting time.

**Solution:** Given Data:

Cast Iron Block of 75 X 20 X 15 Cm

V = 25 Mpm

L = 500 Mm

T = 4mm

F = 1.5mm/Stroke

Stock To Be Removed = 6mm m = 0.5

Angle=45p

V=LN (l+m)/1000 = 500×N (1+0.5)/1000 N=33.33=34 rpm

Metal removal rate W= FTLN = 1.5×4×500×34

= 102 ×103 Mm3/Min

Number of passes N= stock to be removed/depth of cut 6/4=1.5=2

Total cutting time T=LN/Nf = 200×2/34×1.5 = 7.84 min.

**Ans.** Total cutting time = 7.84 min.

## Shaper Machine Unsolved Numerical

1. If n=rpm of the bull wheel=4, cs =cutting speed=3mm/minute and m= the ratio between return time to cutting time= 5. Then find the value of the length of the cutting stroke (l) in mm.

   **Ans.:** 12.5.

2. If l=5, b=5, t=0.4, v=3 and m= 2. Find the value of s. l=length of the cutting stroke in mm, b=width of the work in mm, s=feed in mm, v=cutting speed in mm/minute, m=the ratio between return time to cutting time and t=machining time in a minute.

   **Ans.:** 0.00625.

3. If t=0.10, b=5, s=4, v=1 and m= 2. Find the value of l. l=length of the cutting stroke in mm, b=width of the work in mm, s=feed in mm, v=cutting speed in mm/minute, m=the ratio between return time to cutting time and t=machining time in a minute.

   **Ans.:** 26.66

4. If n=rpm of the bull wheel, cs =cutting speed=3.2mm/minute, and m= the ratio between return time to cutting time= 3 and length of cutting stroke (l)=8mm. Find the value of n.

   **Ans.:** 10

5. If n=rpm of the bull wheel=2, l= length of the cutting stroke in mm=2.5mm and cutting speed = 0.5 mm/minute. m= the ratio between return time to cutting time. Then find the value of m.

   **Ans.:** 9

# 9

# Planer Machine

## Introduction

A planer machine is a commercial machine tool hired for transforming the metal blocks or metal workpiece into the desired calculated shapes by generating the flat surfaces on a workpiece. Similar to a shaper machine, a planer machine is used for forming flat surfaces on the work piece using the single-point cutting tool.

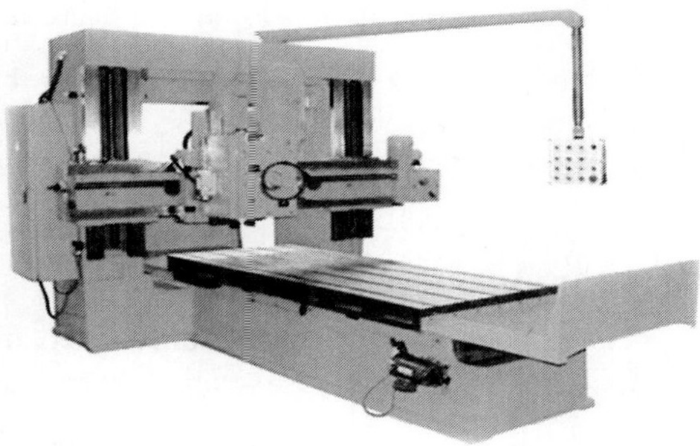

**Figure 302.** Planer Machine

Planer machines are generally hired for operation on the larger workpiece or metal structure (blocks) that cannot be achieved by the shaper machine. They generally have larger beds and larger reciprocating movements covering a greater working length and surface area as compared to shaper machines.

## Parts of Planer Machine

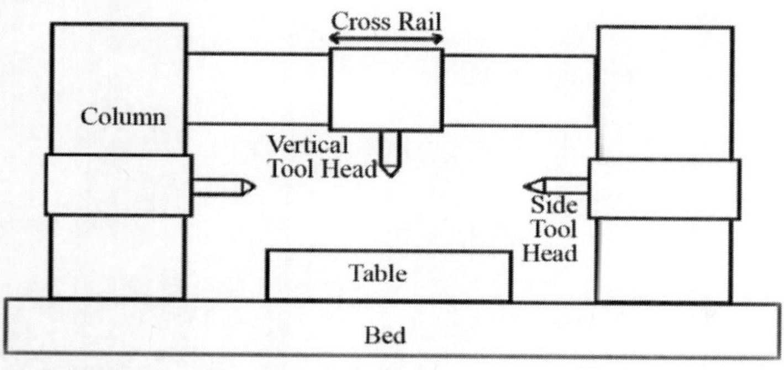


**Figure 303.** Parts of Planer Machine

1. **Bed:** The bed of the planer machine is a horizontal member and the bottommost component of the planer machine. It supports the table and hence the workpiece which reciprocates on the guideways of the bed. Its extended long length orientates the greater strokes of the workpiece and hence helps to achieve the longer or larger workpiece to be shaped in the desired shape.

2. **Column:** The column of the planer machine is the vertical member that houses the feeding mechanism of the machine, and it is also the central controlling unit for the mechanism, feeds, speed and length of the strokes. It also has guideways and a mechanism to adjust the vertical alignment of the vertical member on which the tool heads are fixed.

3. **Table:** The table is a horizontal member that rests and reciprocates on the bed on which the workpiece rests. The reciprocating movement of the bed is responsible for conceiving the feed through the cutting tools.

4. **Crossrail:** The cross rail is the vertical beam-like structure perpendicularly attached to the column of the machine.

   The cross rail has guideways and a mechanism for adjusting the tool heads horizontally or crosswise to align them as per the requirements.

5. **Tool Head:** The planer machine is specifically designed for machining multiple feeds simultaneously at a time and hence it is more productive than the shaper machine.

   The planer machine has multiple tool heads which feed the workpiece simultaneously and individual tool heads can be adjusted according to the desired design of the workpiece. The multiple tool heads can feed the horizontal as well as the vertical surface of the workpiece at the same time during the single reciprocating stroke.

### Types of Planar Machine

The following are the types of planer machine:

1. Double housing planer machine.
2. Pit planer machine.
3. Open side machine.
4. Edge planer machine.
5. Divided table planer machine.

## Double Housing Planer Machine

Two vertical housings are establishing housing: One on every aspect and those are linked on the pinnacle via way of means of a passing member. It has a horizontal pass rail that contains device heads that slide over the vertical faces of the gadget housing.

It has an excessive diploma of floor finish. Work is established on a desk that reciprocates at the same time as the device is hung on the gadget frame. It could make deep cuts and heavy feeds may be implemented to complete the paintings in a quick time. The device is stationary and the paintings are moving. Heavier, stronger, and large equipment are used. Throughout the stroke reduction and return, speeds are uniform.

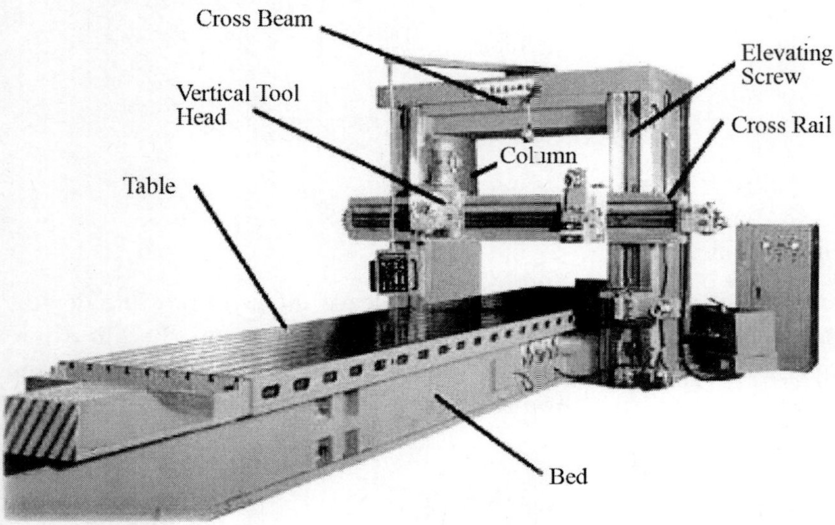


**Figure 304.** Double Housing Planer Machine

## Pit Planer Machine

The desk of the planer is leveled with the floor, so heavy paintings may be loaded. It has device heads and those may be moved horizontally and vertically to have the feed. By method of a motor, a using screw is used for using the column.

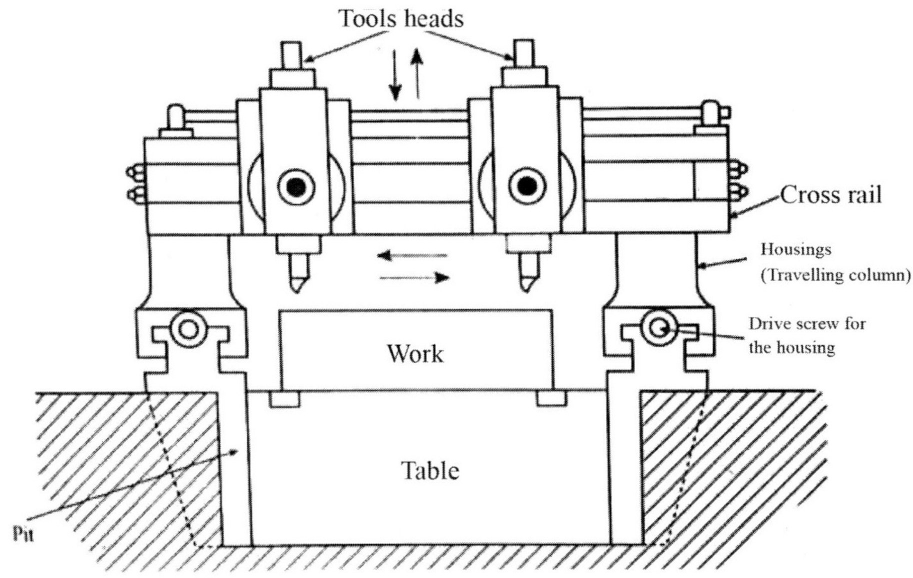


**Figure 305.** Pit Planer Machine

## Open Side Machine

The open facet planer device has 3 device heads hooked up at the device. Single housing will undergo the whole load; therefore, it needs to be inflexible and strong to stand the forces. It can slide alongside the guideways of the housing inside the vertical path which contains the device heads.

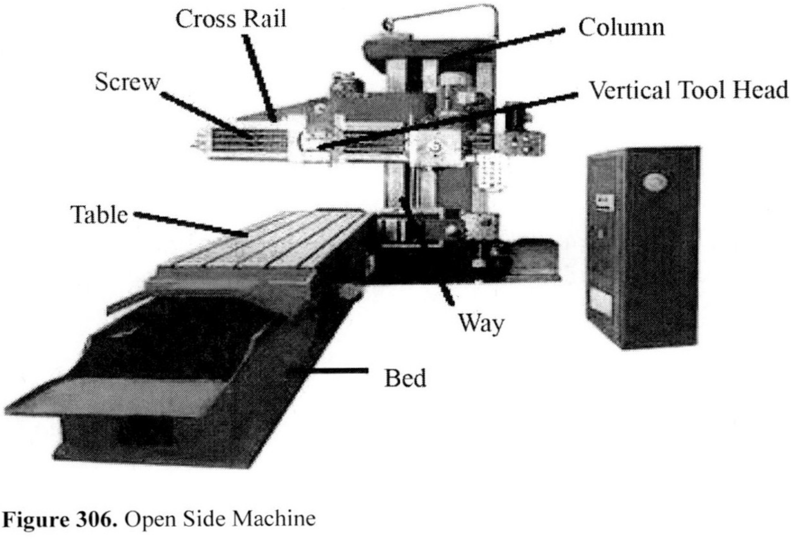


**Figure 306.** Open Side Machine

## Edge Planer Machine

The desk holds the paintings that stay in a desk-bound position. The workpiece may be connected through air-operated clamps. The device head that mounts at the carriage actions alongside horizontal guideways.

## Divided Table Planer Machine

The paintings on one of the tables are loaded, and the opposite element can reciprocate the slicing device for the completing process. Finishing the paintings may be performed after the desk is stopped and the completed process is ejected by moving the desk to the end. Heavy and huge jobs are clamped collectively consequently given the reciprocating motion through the device.

Difference between shaping and planing

| Shaping | Planing |
|---|---|
| 1. In shaping, workpiece is held stationary while cutting tool (ram) is reciprocated across the work. | 1. In planing, cutting tool remains stationary while workpiece (worktable) is reciprocated under the cutter. |
| 2. Here quick-return mechanism is integrated with the ram that holds the cutter. | 2. Here quick- return mechanism is integrated with the worktable that holds the workpiece. |
| 3. Workpiece (bed) imparts feed motion, while cutting tool gives cutting motion. | 3. Workpiece (table) imparts cutting motion, while cutting tool gives feed motion. |
| 4. Shaper is traditionally a small machine and preferred for smaller jobs. | 4. Planer is larger machine and can accommdate heavier and larger jobs. |

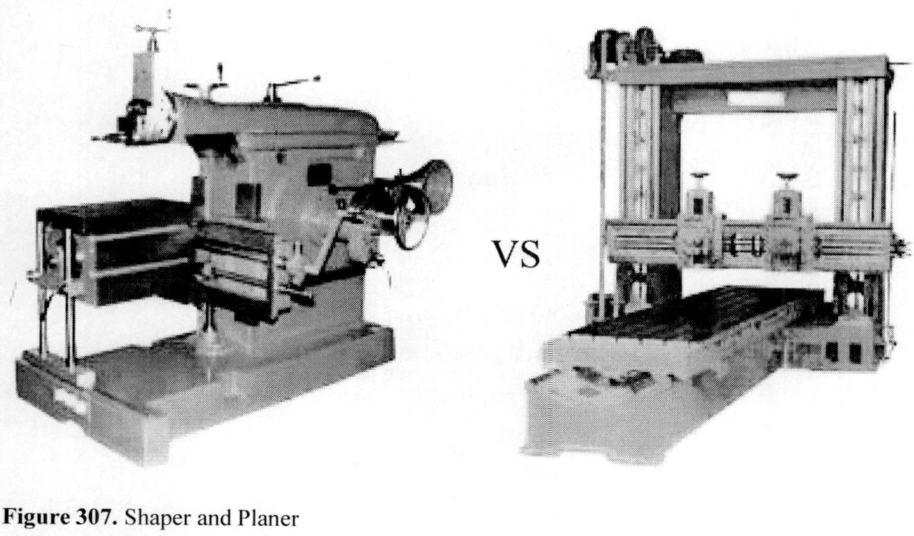


**Figure 307.** Shaper and Planer

## Experiment 1

**Aim:** To perform vertical surfaces operations with Planer Machine

**Materials Required:** Planer machine, planer table, workpiece etc.

**Theory:** The planer machine has multiple tool heads which feeds the workpiece simultaneously and individual tool head can be adjusted according to the desired design of the workpiece. The multiple tool heads can feed the angular, horizontal as well as vertical surface of the workpiece at the same time during the single reciprocating stroke and hence provides a better surface finish and allow for faster work to achieve a flat surface as per the requirement.

### Procedure

1. This operation is done on a planer machine in which the workpiece is planned when the saddle is adjusted horizontally along the direction of the cross rail.
2. Then the tool is set in a position where it gives the desired depth to the cut on the surface.
3. In the planer table, the vertical slide is adjusted perpendicularly.
4. During the operation, a return stroke of the apron is swiveled that's why the tool will correctly come out of the machine.

**Result:** Vertical surfaces are obtained.

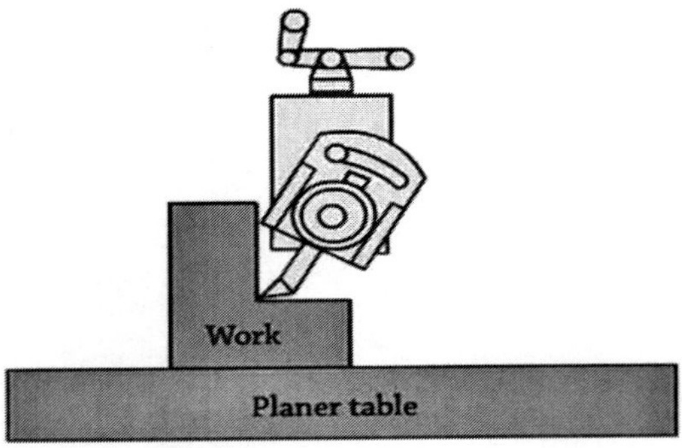


**Figure 308.** Planer Machine

### Precautions

1. Work under the supervision of an expert
2. Maintain a safe distance from a machine during the operation
3. Work with the proper safety equipment.
4. Wear all safety things and dress
5. Not move anywhere leaving the machine running.
6. Go through all the instructions, processes, and steps once again properly.

## Experiment 2

**Aim:** To perform the horizontal surface operations with Planer Machine.

### Materials Required

- Planer machine
- Workpiece,
- Clapper block

**Theory:** Horizontal planer operations on the workpiece are performed on the planner when the workpiece reciprocates parallel to the table and the tool is aligned on the horizontal cross-rail guideways. Properly fix the work on the table then.

### Procedure

1. Set the cutting speed as per the requirement.
2. The feed of the tool is given as required.
3. Give suitable depth of cut.
4. For the rough cuts give a suitable depth for a cut.
5. Then finish the job by giving less depth of cut.

**Result:** The desired horizontal surface is obtained.

### Precautions

1. Work under the supervision of an expert
2. Maintain a safe distance from a machine during the operation
3. Work with the proper safety equipment.
4. Wear all safety things and dress
5. Not move anywhere leaving the machine running.

## Experiment 3

**Aim:** To perform angular surface operations with Planer Machine.

**Materials Required:** Planer machine, workpiece etc.

**Theory:** The planer machine has multiple tool heads which feeds the workpiece simultaneously and individual tool head can be adjusted according to the desired design of the workpiece. The multiple tool heads can feed the angular, horizontal as well as vertical surface of the workpiece at the same time during the single reciprocating stroke and hence provides a better surface finish and allow for faster work to achieve a flat surface as per the requirement.

### Procedure

1. Planning angular surfaces means to obtain dovetails & V grooves.
2. Firstly, held the work on the table.
3. Then according to the required angle Swivel the tool head.
4. Then set the apron away from work during the machining.
5. Now as per the requirement give down the feed.

   **Result:** The desired angular surface is obtained.

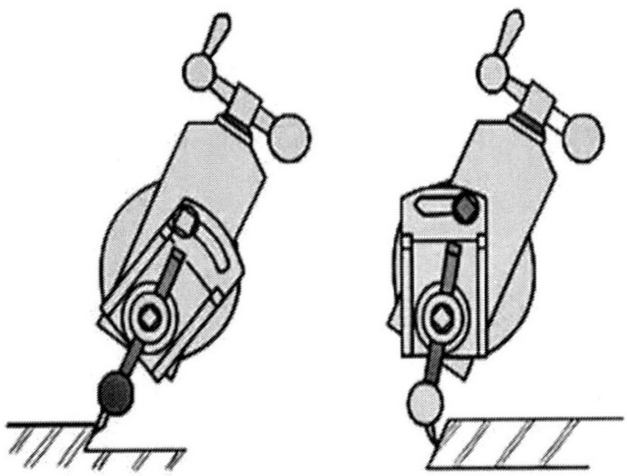

**Figure 309.** Planer Machine

### Precautions

1. Work under the supervision of an expert
2. Maintain a safe distance from a machine during the operation
3. Work with the proper safety equipment.
4. Wear all safety things and dress
5. Not move anywhere leaving the machine running.
6. Go through all the instructions, processes, and steps once again properly.

## Experiment 4

**Aim:** To perform the formed surface operations with Planer Machine

**Materials Required:** Planer machine, workpiece etc.

**Theory:** The planer machine has multiple tool heads which feeds the workpiece simultaneously and individual tool head can be adjusted according to the desired design of the workpiece. The multiple tool heads can feed the angular, horizontal as well as vertical surface of the workpiece at the same time during the single reciprocating stroke and hence provides a better surface finish and allow for faster work to achieve a flat surface as per the requirement.

### Procedure

1. Firstly, a square nose tool is fixed up in the tool head.
2. Feeding the tool in both horizontal and vertical directions simultaneously the required form of the workpiece is obtained.
3. Then provide a suitable depth of cut.
4. Then set the apron away from work.
5. By adding a special fixture this work can also be done.

   **Result:** Desired surfaces are obtained.

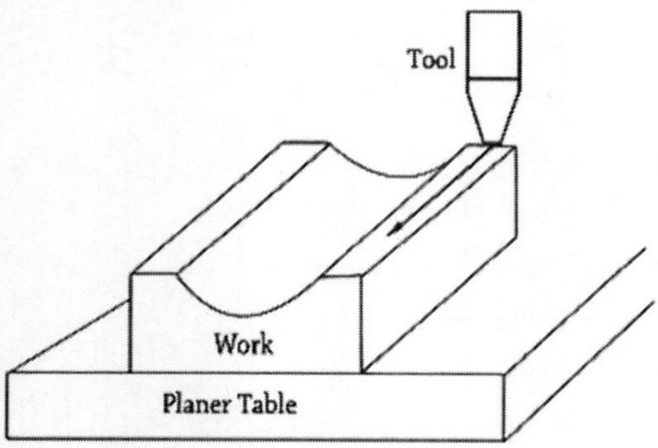


**Figure 310.** Planing Formed Surfaces

### Precautions

1. Work under the supervision of an expert
2. Maintain a safe distance from a machine during the operation
3. Work with the proper safety equipment.
4. Wear all safety things and dress
5. Not move anywhere leaving the machine running.
6. Go through all the instructions, processes, and steps once again properly.

## Planer Machine Objective Questions

1. In a planer machine, the cutting tool moves:
   a) Vertically  b) **Horizontally**
   c) Diagonally  d) In all three directions
2. The primary function of a planer machine is to:
   a) Create cylindrical holes  b) Mill complex shapes
   c) **Produce flat surfaces**  d) Generate internal threads
3. The reciprocating motion in a planer is usually achieved by:
   a) Gear mechanism  b) Hydraulic system
   c) Pneumatic system  d) **Crank and connecting rod mechanism**
4. The workpiece is usually secured to the bed of the planer using:
   a) **Clamps**  b) Chucks
   c) Collets  d) Vises
5. Which of the following is a type of planer machine that specializes in machining the top surface of large workpieces?
   a) Surface planer  b) Double housing planer
   c) Open side planer  d) **Plano-miller**
6. The stroke length of a planer machine refers to:
   a) **The distance the tool moves in one cutting pass**
   b) The maximum depth of cut
   c) The distance between the tool and the workpiece
   d) The length of the machine bed
7. The cross-rail adjustment in a planer machine is used to control:
   a) **The vertical position of the tool**
   b) The horizontal position of the tool
   c) The rotational speed of the tool
   d) The feed rate of the tool
8. Which type of planer machine is known for its heavy-duty capabilities and is used for machining large workpieces?
   a) Open side planer  b) **Double housing planer**
   c) Surface planer  d) Plano-miller

9. The tool head of a planer machine is mounted on:
   a) The cross-rail b) The bed
   c) The column d) **The ram**
10. Planer machines are commonly used for:
   a) High-speed cutting b) Precision drilling
   c) **Heavy-duty roughing** d) Micro-machining

## Planer Machine Subjective Questions

1. Write down the application of a planner.
2. What are the major components of a planner?
3. Write about various tool supports for a planner.
4. Write about various clamping methods of works through sketch.

## Planer Machine Solved Numericals

Q1 A double-housing planer machine has a stroke length of 600 mm. If the cutting speed is 15 m/min and the number of strokes per minute is 30, calculate the material removal rate in cubic centimeters per minute.

**Solution:**

Material Removal Rate (MRR) = Stroke Length × Cutting Speed × Width of Cut × Depth of Cut × Strokes per Minute

Given:

Stroke Length = 600 mm

Cutting Speed = 15 m/min

Width of Cut (W) = Let's assume 150 mm

Depth of Cut (D) = Let's assume 3 mm

Strokes per Minute = 30

MRR = 600 mm × (15 m/min) × (150 mm) × (3 mm) × 30 = 2,025,000 cubic mm/min = 202.5 cubic cm/min

Q2 A planer is machining a workpiece at a feed rate of 0.4 mm/stroke. If the length of the workpiece is 800 mm and each return stroke takes 5 seconds, calculate the machining time.

**Solution:**

Machining Time = (Total Length / Feed Rate) + (Return Stroke Time × Number of Return Strokes)

Given:

Feed Rate = 0.4 mm/stroke

Total Length = 800 mm

Return Stroke Time = 5 seconds

Number of Return Strokes = 1 (since one return stroke is made per machining cycle)

Machining Time = (800 mm / 0.4 mm/stroke) + (5 seconds × 1) = 2000 strokes + 5 seconds = 2005 seconds

Q3 A planer is used to machine a workpiece having a width of 250 mm. If the feed rate is 0.25 mm/stroke and each pass removes 1.5 mm of material, calculate the number of passes required to complete the machining.

**Solution:**

Number of Passes = Depth of Cut / Width of Cut

Given:

Depth of Cut = 1.5 mm

Width of Cut = 250 mm

Number of Passes = 1.5 mm / 250 mm = 0.006 passes

Since we can't have a fraction of a pass, you would typically round up to the nearest whole number. Therefore, the number of passes required is 1 pass.

Q4 In a planer machine, the cutting tool travels at a speed of 12 meters per minute during the cutting stroke. If the stroke length is 450 mm and each pass removes 0.8 mm of material, calculate the time taken to complete one machining cycle.

**Solution:**

Time for One Machining Cycle = (Cutting Stroke Time) + (Return Stroke Time)

Given:

Cutting Stroke Speed = 12 m/min

Stroke Length = 450 mm

Depth of Cut = 0.8 mm

Return Stroke Speed = Assuming the same as cutting stroke speed

Cutting Stroke Time = Stroke Length / Cutting Stroke Speed

Return Stroke Time = Stroke Length / Return Stroke Speed

Cutting Stroke Time = 450 mm / (12 m/min) = 37.5 minutes

Return Stroke Time = 450 mm / (12 m/min) = 37.5 minutes

Time for One Machining Cycle = 37.5 minutes + 37.5 minutes = 75 minutes

Q5 A planer machine is operating with a stroke length of 600 mm, a cutting speed of 20 m/min, and a feed rate of 0.3 mm/stroke. If each pass removes 1.2 mm of material, calculate the time taken to complete the machining of a workpiece that is 800 mm long.

**Solution:**

Total Machining Time = (Number of Passes × Time for One Pass) + (Number of Return Strokes × Return Stroke Time)

Given:

Stroke Length = 600 mm

Cutting Speed = 20 m/min

Feed Rate = 0.3 mm/stroke

Depth of Cut = 1.2 mm

Total Length = 800 mm

Number of Passes = Depth of Cut / Feed Rate

Time for One Pass = Stroke Length / Cutting Speed

Return Stroke Time = Stroke Length / Return Stroke Speed

Number of Passes = 1.2 mm / 0.3 mm/stroke = 4 passes

Time for One Pass = 600 mm / 20 m/min = 30 minutes

Return Stroke Time = 600 mm / (Assuming the same return stroke speed) = 30 minutes

Total Machining Time = (4 passes × 30 minutes/pass) + (1 return stroke × 30 minutes) = 120 minutes + 30 minutes = 150 minutes

## Planer Machine Unsolved Numericals

1. A planer machine is used to machine a workpiece with a stroke length of 600 mm. If the cutting speed is 18 m/min and the feed rate is 0.25 mm/stroke, calculate the time required for a single pass.

   **Ans.:** Time = 600 mm / (18 m/min * 1000 mm/m) = 0.0333 minutes or approximately 2 seconds.

2. A planer machine is used to machine a workpiece with a length of 800 mm. The machine operates with a stroke speed of 20 strokes per minute, and each stroke removes 0.5 mm of material. Calculate the time required to complete the machining operation.

   **Ans.:** Time = (1600 strokes * 3 seconds/stroke) + 0 = 4800 seconds or 80 minutes.

3. A planer machine is used to machine a workpiece with a width of 250 mm. If each pass removes 1 mm of material and the feed is 0.15 mm/stroke, calculate the number of passes required to complete the machining.

   **Ans.:** Number of Passes = 1 mm / 0.15 mm/stroke H" 6.67 passes.

4. A planer machine is used to machine a workpiece with a length of 1200 mm. The cutting speed is 12 m/min, and the feed rate is 0.2 mm/stroke. Calculate the time required to complete the machining if each pass removes 0.5 mm of material.

   **Ans.:** Time = 625 minutes.

5. A planer machine is used to machine a workpiece with a width of 300 mm. If the feed rate is 0.3 mm/stroke and each pass removes 1.2 mm of material, calculate the number of passes required to complete the machining.

   **Ans.:** Number of Passes = 1.2 mm / 0.3 mm/stroke = 4 passes.

# 10

# Milling Machine

## Introduction

The milling machine is one of the maximum flexible conventional system equipment with a huge variety of steel-reducing capabilities. Many complex operations along with indexing, multi-milling and saddle milling, etc. may be completed on a milling system. A milling device is a tool that cuts steel at the same time as the workpiece is fed closer to a rotating multi-element milling cutter. This machine can also contain one or greater knives at a time. That is why a milling device has extensive utility in production work. It's better for other machines when it comes to precision and a better surface finish. It is also designed to perform a variety of jobs in the tool room.

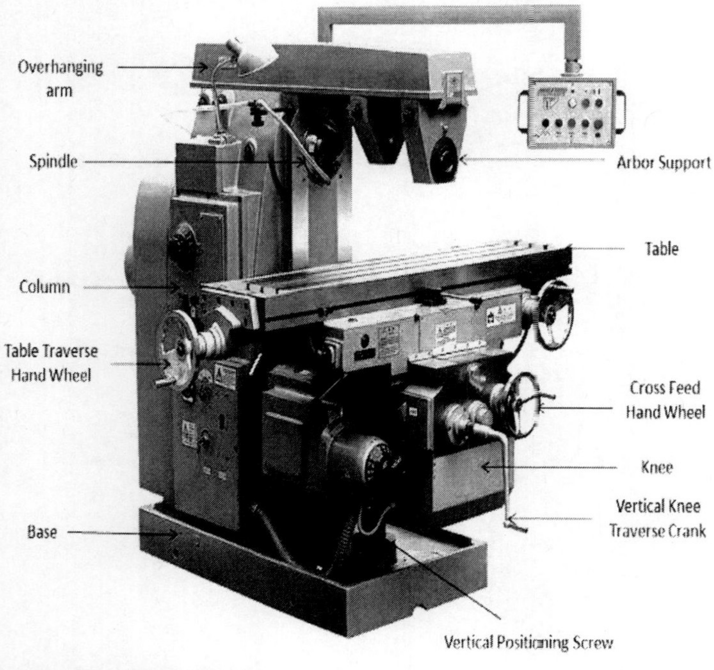


**Figure 311.** Milling Machine

Milling machines also called multitasking machines (MTM) which can be flexible machines able to milling and turning even materials. The milling device is prepared with a milling cutter which enables cast-off fabric from the workpiece surface.

## Parts of the Milling Machine

### Main Parts of Milling Machine are

1. **Base:** It takes a load of all other parts. The whole machine is being mounted on base. It is also a reservoir for coolant. Mostly it is made up of cast iron, to make it rigid.
2. **Column:** It consists of a driving mechanism and motor. It is a main supporting frame. The top of a column over the arm is supported.
3. **Knee:** The knee is attached to the column. With the motion of the knee, the space between the cutter and workpiece may be adjusted.
4. **Table:** It top surface consists of a longitudinal T-slot and accommodates clamping bolts for fixing the workpiece.

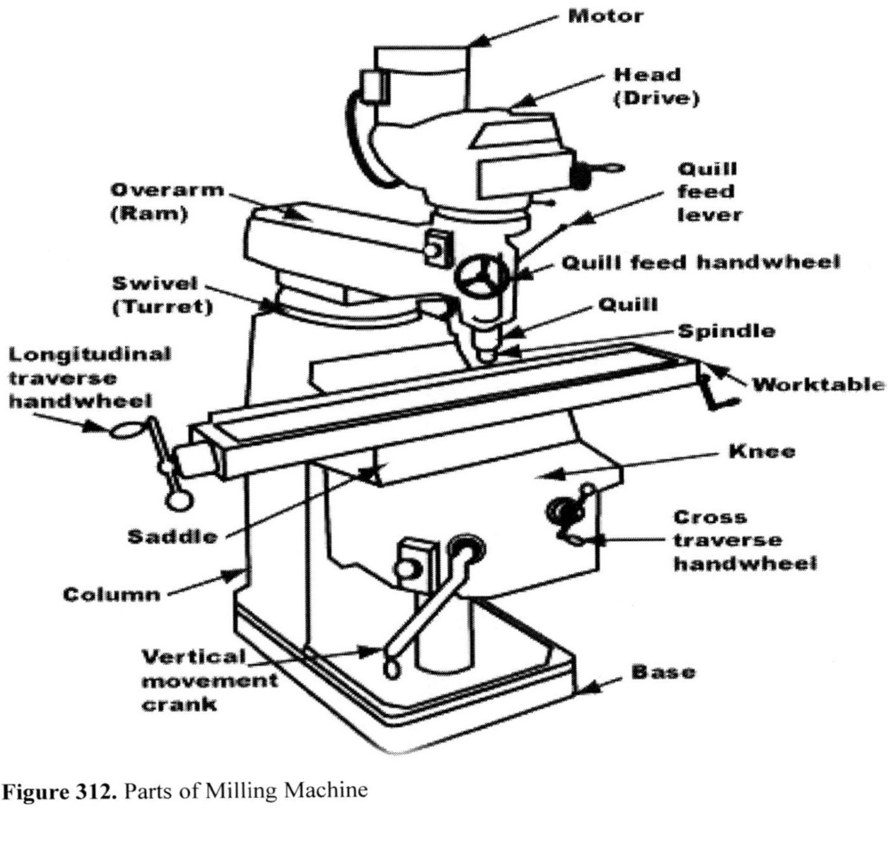


**Figure 312.** Parts of Milling Machine

5. **Saddle:** It is an intermediate part between the table and the knee. It gives support to the table. A saddle can be adjusted crosswise along with the guideways on the top of the knee to provide cross-feed to the table.
6. **Over Arm:** It is the heavy support that is provided at the top of the milling machine. It supports an arbor.
7. **Spindle or Arbor:** Arbor holds the milling cutters during the operation. It is of two types - standard arbor and stub arbor.

## Types of Milling Machine

### 1. Knee and Column Type

i) **Horizontal or Planer Milling Machine:** It is used for heavy job work. It has a horizontal spindle so it is called a horizontal milling machine. The table can be fed in longitudinal, vertical, or cross directions.

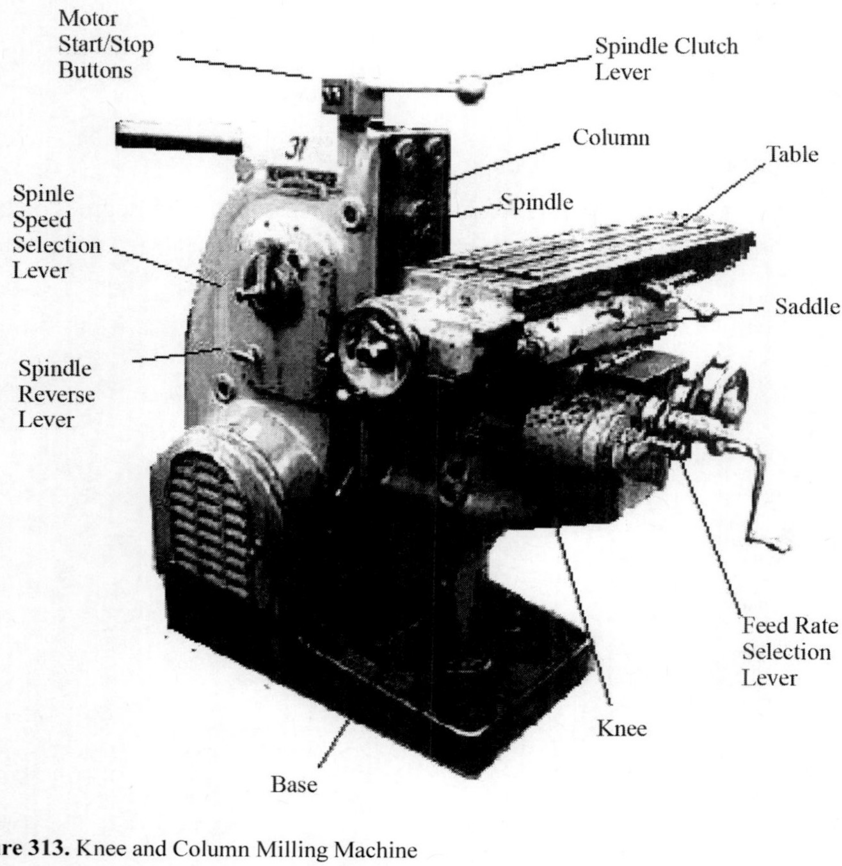


**Figure 313.** Knee and Column Milling Machine

ii) **Vertical Milling Machine:** In this type the spindle is perpendicular to the table. The milling cutter is fixed on a spindle and rotated about its own axis. Vertical milling machine is subcategorized as bed milling and turret milling.

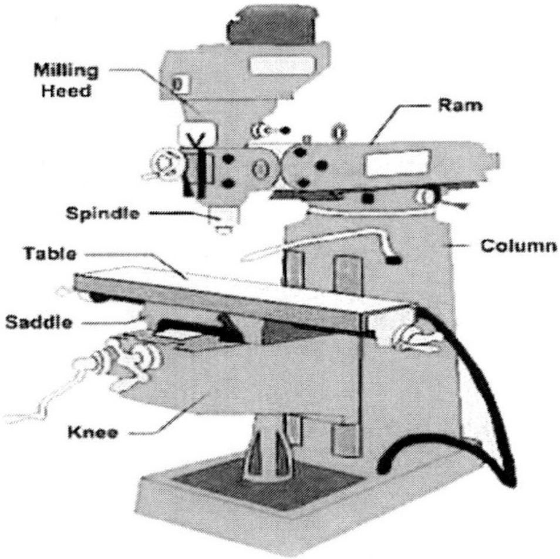


**Figure 314.** Vertical Milling Machine

## 2. Universal Milling Machine

It can perform various operations. It can pivot at an angle of 45 degrees on both sides. Other than movement in three directions, it can also perform helical operations.

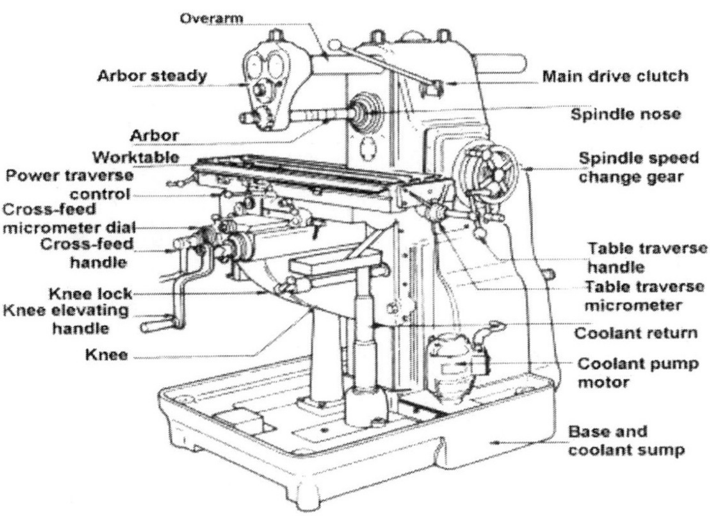


**Figure 315.** Horizontal Milling Machine

### 3. Fixed Bed Type

It is mounted on a rigid bed. It provides only longitudinal movement. These are high-production milling machines.

i) **Simple Milling Machine:** In this type spindle head can move in a single direction.

ii) **Duplex Milling Machine:** It has an adjustable horizontal spindle head attached to row separate sides in which vertical columns are on each side.

iii) **Triplex Milling Machine:** It has two horizontal spindle heads attached to two separate vertical columns and an additional and additional vertical spindle. It can drive in all three directions.

### 4. Planer Type Milling Machine

This type is used for heavy work. In this maximum four tool heads can be mounted and adjust these tools heads in transverse and vertical directions. It is also called Plano millers.

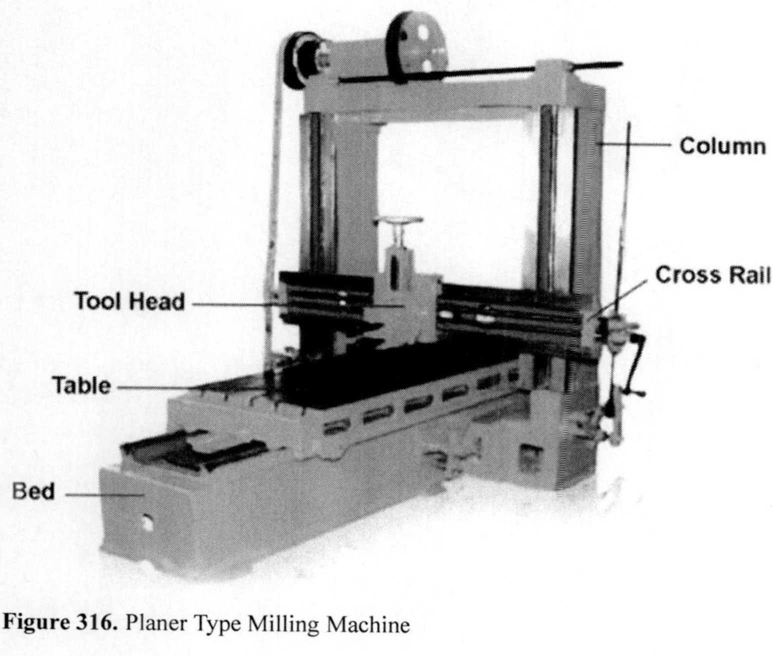


**Figure 316.** Planer Type Milling Machine

## 5. Production Milling Machine

i) **Rotatory Table Milling Machine:** It consists of a circular table that rotates on a vertical axis. It is used for mass production. It gives three times more production than Plano mills since machining is continuous.

ii) **Drum Type Milling Machine:** In this type, two ends of the workpiece can be machined simultaneously. It has a drum and along the periphery of the drum, the fixtures are mounted to support the workpiece.

iii) **Tracer Controlled Milling Machine:** It is used in the production of dies used in presswork, die-casting, and forging. It is used to produce shapes in a workpiece.

iv) **CNC Milling Machine:** It is an upgraded version of bed type milling machine. In this movements are controlled hydraulically using computers.

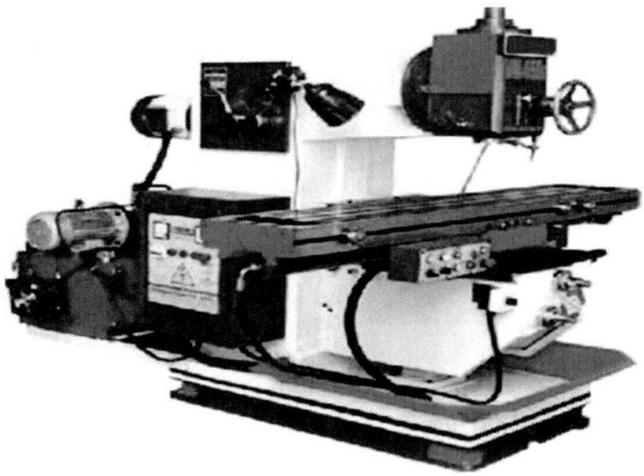

**Figure 317.** Production Milling Machine

## Cutting Tools

### For Horizontal Milling Machine

i) **Slab Mills:** For heavy reduction of massive and flat surfaces.

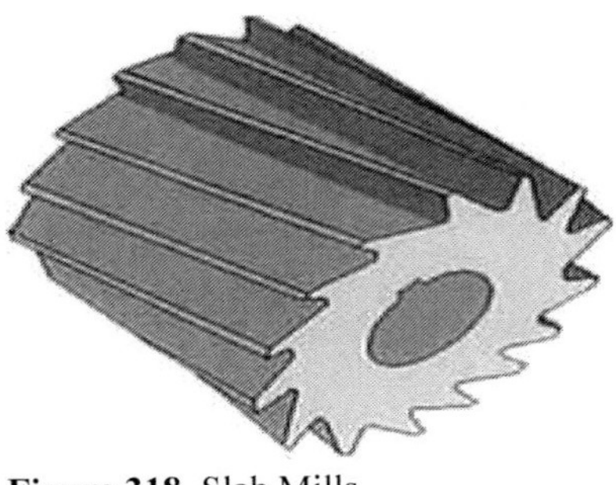

**Figure 318.** Slab Mills

ii) **Side and Face:** Cutters this sort of cutter has reducing edges at the outer edge and aspects of the enamel for reducing shoulders and slots.

**Figure 319.** Side and Face

iii) **Slitting Saw:** For slicing deep slots or for parting off.

**Figure 320.** Slitting Saw

### For Vertical Milling Machine

i) **End Mills:** It is used for facing, slotting and profile milling

**Figure 321.** End Mills

ii) **Rough Cut End Mills:** For fast steel removal.

**Figure 322.** Rough Cut End Mills

iii) **Slot Drills:** For generating pockets without drilling a hollow earlier than hand.

**Figure 323.** Slot Drills

iv) **Face Milling**: Used for heavy cutting.

**Figure 324.** Face Milling

## Milling Process

In this process, the steel pieces are removed with the help of a rotating cutter having one or additional cutting teeth. Cutting motion is achieved through the manner of feeding the workpiece in competition to the rotating cutter.

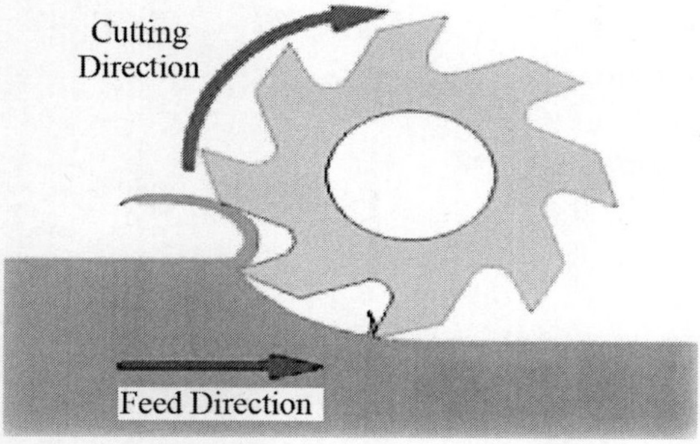


**Figure 325.** Milling Process

**Spindle Speed:** Spindle velocity in revolution consistent with minute (R.P.M.) for the cutter may be calculated from the equation:

$$N = \frac{CS \times 1000}{\pi d}$$

Where,

N = R.P.M. of the cutter CS = Linear Cutting Speed of the cloth in m/min. (see desk 1) d = Diameter of cutter in mm.

**Feed Rate:** Feed charge (F) is described because of the charge of a journey of the workpiece in mm/min. But maximum device providers propose it because of the motion consistent with the teeth of the cutter (f). Thus,

$F = f.u.N$

Where,

C.S. and feed charge for a few not unusual place materials:

| Tool Material<br>Material | High Speed Steel<br>Cutting Speed | Feed (f) | Carbide<br>Cutting Speed | Feed (f) |
|---|---|---|---|---|
| Mild Steel | 25 | 0.08 | 100 | 0.15 |
| Aluminium | 100 | 0.15 | 500 | 0.3 |
| Hardened Steel | — | — | 50 | 0.1 |

**Depth of cut:** Depth of reduction is immediately associated with the performance of the reducing process. The deeper the reduction the quicker might be the manufacturing rate. Yet, it nonetheless relies upon the power of the cutter and the cloth to be reduced.

For a sure kind of cutter, an ordinary variety of reducing might be endorsed through the supplier. Nevertheless, it must be cited that a finer reduction is commonly related to a higher floor end in addition to extended device life.

## Direction of Cutter Rotation

1. **Up-Cut Milling:** In up reduce milling, the cutter rotates in a course contrary to the desk feed. It is conventionally utilized in maximum milling operations due to the fact the backlash among the lead screw and the nut of the gadget desk may be eliminated.

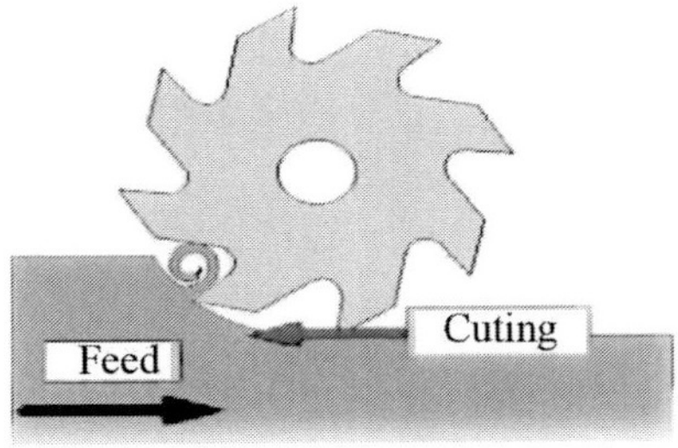


**Figure 326.** Up Cut Milling

2. **Down Cut Milling:** In down cut edge, the cutter rotates with among a similar path as a result of the actual fact of the table feed. This technique, while nicely treated, would need a lot less electricity in feeding the table and provide the next floor finish at the piece of work.

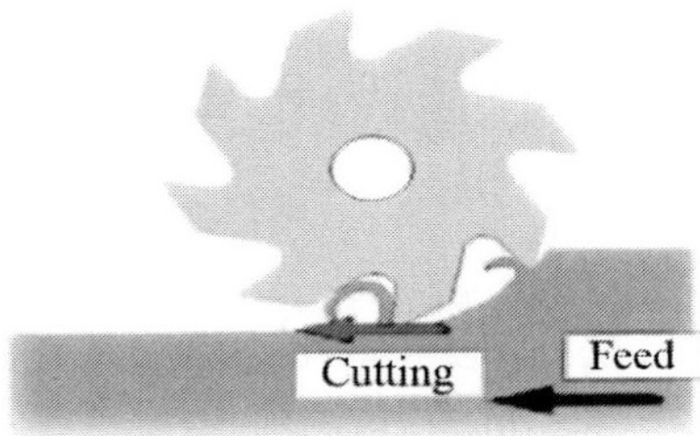


**Figure 327.** Down Cut Milling

## Milling Machine Operations

1. Plain Milling Operation.
2. Face Milling Operation.
3. End Milling Operation.
4. T-slot Milling Operation.
5. Gear Milling Operation.
6. Grooves Milling Operation.
7. Form Milling Operation.
8. Side milling.
9. Straddle milling.
10. Angular milling.
11. Gang milling.
12. Profile milling.
13. Sawmilling.
14. Helical milling.

## 1. Plain Milling Operation

i) The undeniable milling is the maximum easy sort of milling tool operation.

ii) Provide an exact, flat and horizontal ground which is parallel to the axis of rotation of a definite cutter.

iii) The operation is also referred to as slab milling.

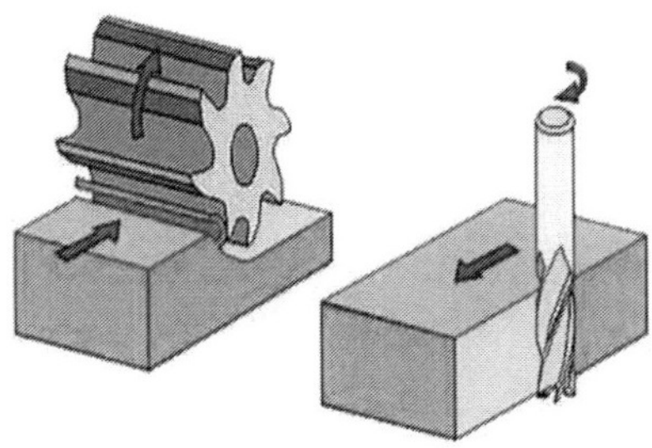

**Figure 328.** Plain Milling Operation

iv) By rotating the vertical feed screw of the table depth of the cut can be obtained. And also the tool begins once selecting the proper speed and feed.

## 2. Face Milling Operation

i) Face milling is the handiest milling system operation.

ii) This operation is finished with the help of employing a face edge cutter gyrated more or less associate degree axis perpendicular to the surface.

iii) The cutter is positioned on a stub arbor to make a flat surface, this process is done in plain milling.

iv) The intensity of reduction is adjusted with the resource of the usage of rotating the cross-feed screw of the table.

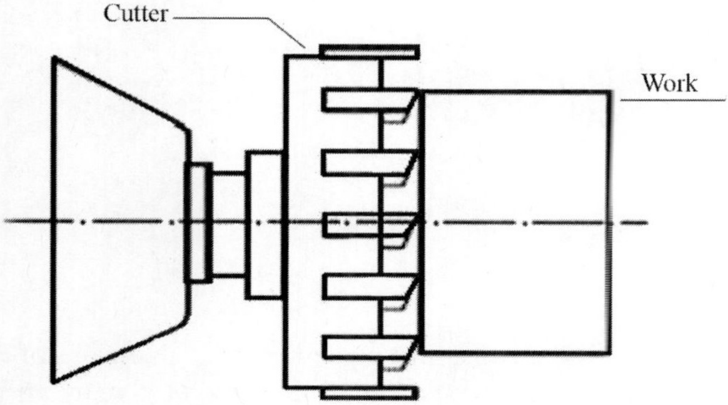


**Figure 329.** Face Milling Operation

## 3. End Milling Operation

i) In this operation, we produce a flat surface that is horizontal, and vertical and forms an angle to the table.

ii) A vertical milling device is extra appropriate for end milling operation.

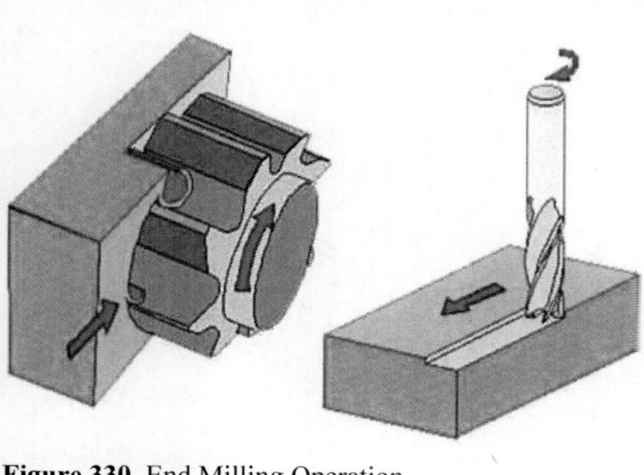

**Figure 330.** End Milling Operation

## 4. T-Slot Milling Operation

In this firstly a plain slot is cut on the workpiece by a side and face milling cutter then a T-slot cutter is fed from the end of the workpiece.

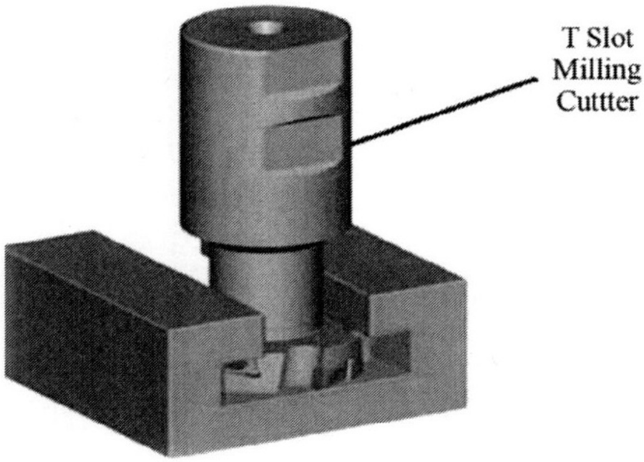


**Figure 331.** T-slot Milling Operation

## 5. Gear Milling Operation

i) The gear slicing operation is done in a milling device through the use of a form- relieved cutter. The cutter can be a cylindrical kind or give up mill kind.

ii) The cutter profile suits precisely the enamel area of the tools.

iii) Equally spaced system teeth are reduced on a system clean through keeping the artwork on a commonplace area diving head after which indexing it.

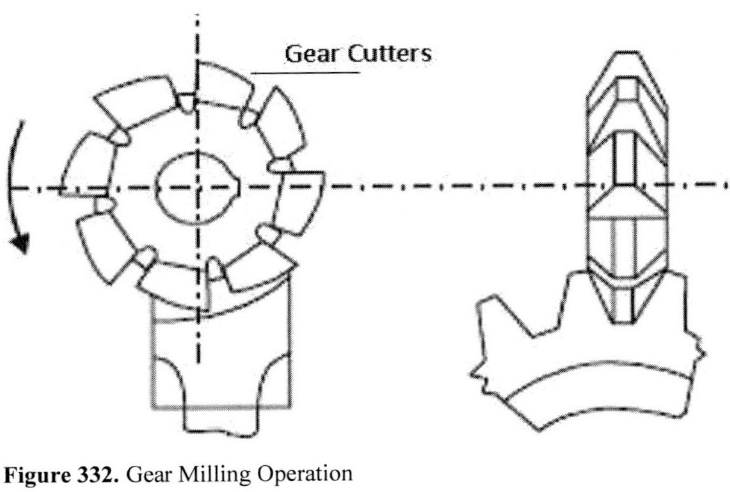


**Figure 332.** Gear Milling Operation

## 6. Grooves Milling Operation

This operation is used to make various types of the groove by using a plain milling cutter. They form open slots.

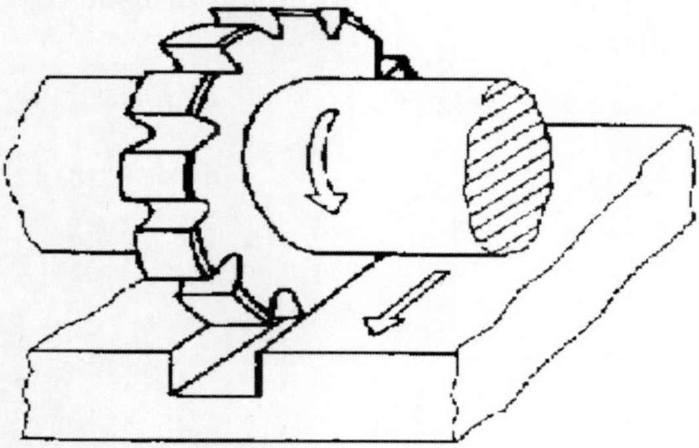

**Figure 333.** Grooves Milling Operation

## 7. Form Milling Operation

It is the operation of producing the irregular contour with the useful resource of the use of using shape cutters.

The abnormal shape can be convex, concave, or of another

form. After machining, the fashioned floor is inspected with the aid of using a template gauge.

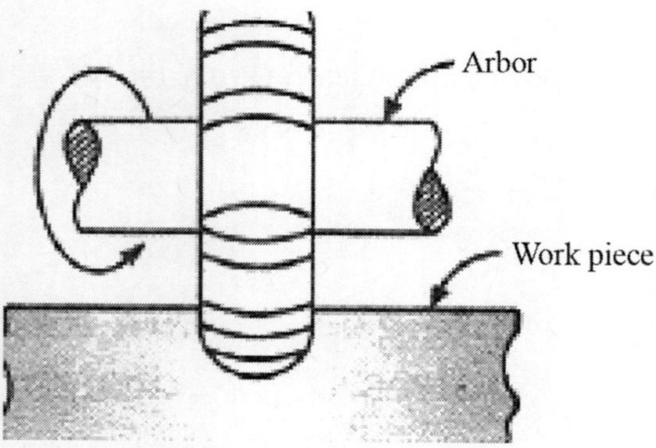


**Figure 334.** Form Milling Operation

## 8. Side Milling

It is a technique of producing flat and vertical surfaces.

## 9. Straddle Milling

i) The use of side milling cutters produces a flat vertical surface on both sides of a workpiece.

ii) The distance between the 2 cutters is adjusted with the aid of using the usage of appropriate spacing collars.

iii) It is used to make square and hexagonal surfaces.

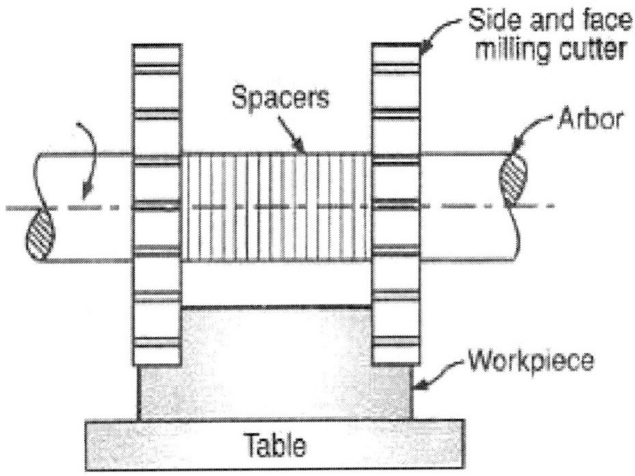


**Figure 335.** Straddle Milling

## 10. Angular Milling

i) In this operation, we produce an angular surface on a workpiece.

ii) One easy instance of angular milling is the manufacturing of V-blocks.

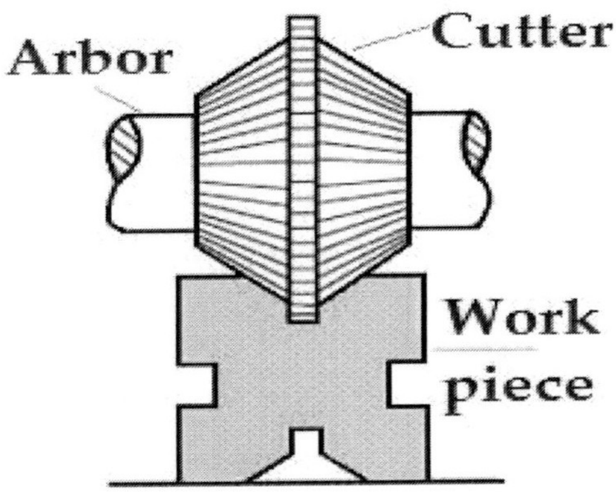


**Figure 336.** Angular Milling

## 11. Gang Milling

i) It is the operation of machining numerous surfaces of a bit of labor on an equal time with the assist of exploitation feeding the desk closer to a range of cutters having an equivalent or specific diameters put in at the arbor of the machine.

ii) The approach saves a whole lot of machining time and is extensively utilized in repetitive work.

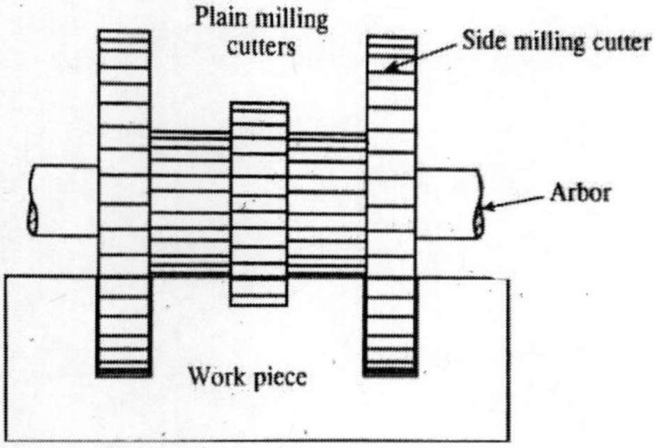


**Figure 337.** Gang Milling

## 12. Profile Milling

i) It is the operation of reproducing complex shapes on the workpiece and outline of a template.

ii) Different types of cutters are used for profile milling.

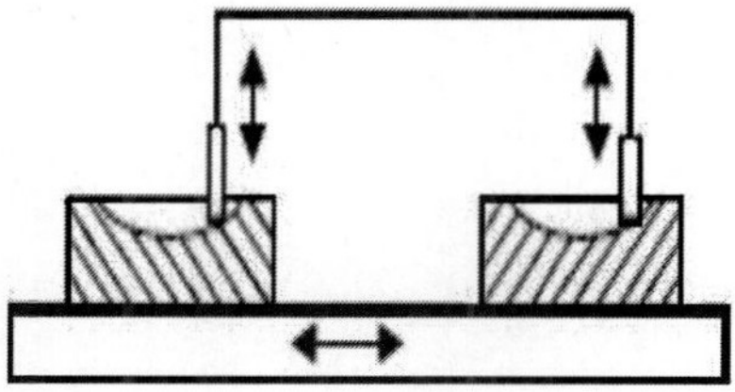

**Figure 338.** Profile Milling

## 13. Saw Milling

i) Saw-milling is the operation of manufacturing slim slots or grooves on a workpiece via way of means of the usage of a saw-milling cutter.

ii) The saw-milling is additionally completed for the whole parting-off operation.

iii) The cutter and therefore the work is set in a very manner in order that the cutter is true away positioned over one in all the T-slots of the table.

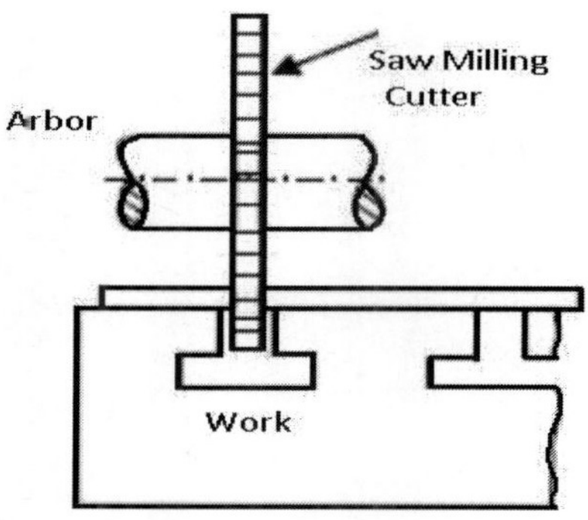


**Figure 339.** Saw Milling

## 14. Helical Milling

i) It is the operation of producing helical flutes or grooves throughout the outer fringe of a cylindrical or conical workpiece.

ii) The operation is executed by rotating the desk to the desired helix angle. And then via rotating and feeding the workpiece in competition to rotary lowering edges of a milling cutter.

iii) Production of the helical milling cutter, helical gears, reducing helical grooves or flutes on a drill clean or a reamer.

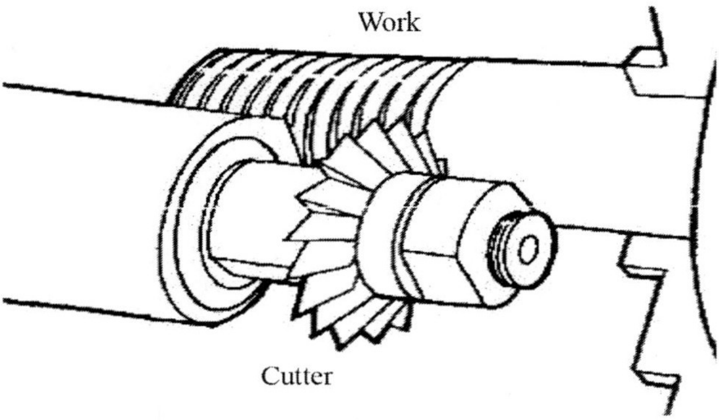


**Figure 340.** Helical Milling

## Experiment 1

**Aim:** Gear tooth forming by the milling machine.

**Materials Required:** Workpiece.

### Tools Required

- Horizontal milling machine.
- Vernier caliper.
- Indexing attachment.
- Spur gear cutter.
- Gear tooth vernier.

### Theory

**Indexing:** It is the technique of dividing the circumference of a round workpiece in similarly spaced divisions.

1. **Types of indexing head:** It is of 3 types as given below:
   i) Plain or simple indexing head.
   ii) Universal dividing head.
   iii) Optically dividing head.
2. **Indexing methods:**
   i) Direct or rapid indexing.
   ii) Simple or plain indexing.
   iii) Compound indexing.
   iv) Differential indexing.
   v) Angular indexing.
3. **Simple or Plain Indexing:** In this method, the workpiece is rotated by turning the crank. Different index plates that have various variety of holes are used to grow the indexing range
4. **Compound Indexing:** This method use two different circles of one plate. Sometimes it is also called the hit and trial method. It involved two separate indexing movements.
5. **Differential Indexing:** This is the same as compound indexing but it is an automatic method.

### Procedure

1. Obtain gear blank using a circular rod, which is casted, in the required dimensions.
2. Drill a hole at the face center of a rod of a given depth.
3. Blank of given width and external diameter is parted off.
4. Fix gear blank between dividing head center and dead center

5. The required index plate is selected and fixed.
6. Adjust elevating screw and raise milling machine to align gear blank.
7. Select the cutting speed and turn the machine on.
8. Gear blank is fed against gear cutter
9. On completing each tooth, the gear blank is removed.
10. The sector arm is moved to another point
11. The crank is rotated to the required number of rotations.
12. Crankpin is fixed at the required index plate hole number to exact no. Of equal division on gear blank.

## Calculation

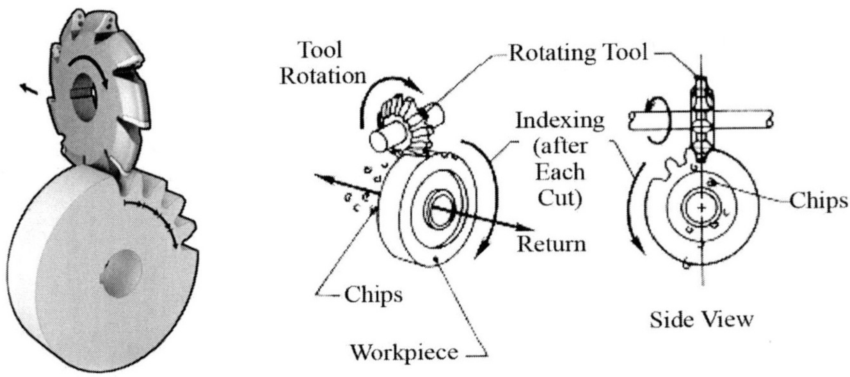


**Figure 341.** Milling Machine

No. of teeth cut on gear, blank is (n)

Blank diameter = (z +2) m

z = no. of teeth

**Result:** The required spur gear is obtained.

Gear Cutting on Milling Machine.

## Precaution

1. Maintain distance from the revolving cutter.
2. Always use a brush to clean chips from the workpiece.
3. Use a splash guard while using the cutting oil.
4. Must wear safety gear before starting the operation.
5. You must have at least one person during the operation for any need arises.
6. Check all equipment before any operation.
7. Don't use a machine at high.

## Experiment 2

**Aim:** To cut the gear on a milling machine and find the depth of cut.

**Materials Required:** Workpiece.

### Tools Required

- Cutting tool.
- Dog carrier.
- Mandrel.

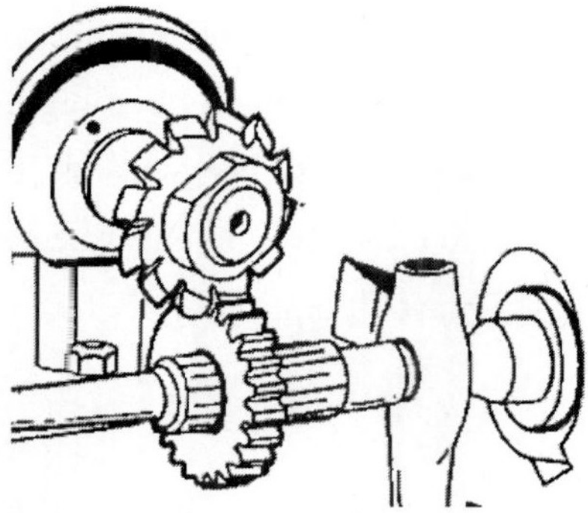

Figure 342. Gear of Milling Machine

### Theory

A milling machine is a device that is based on the principle that removal of chips from workpiece as it fed a rotating cutter. It covers a variety of operations such as face milling operation, slot or slab milling operation, gear cutting operation, end cutting operation etc.

1. **Gearing Cutting Operation**: The process of cutting equally spaced, identical gear teeth on a gear workpiece with the help of a gear cutting tool.
2. **Indexing**: It is the operation of dividing the outer edge of a workpiece into any wide variety of identical parts.

### Procedure

1. To select the raw blank with reference to the number of teeth to be cut.
2. The mandrel is used to mount gear blank.
3. Don't use the machine at high speed.

### Calculation

The depth of the cut D = 2.156/DP mm

Where DP = Diametrical Pitch

**Result:** The depth of the cut is mm.

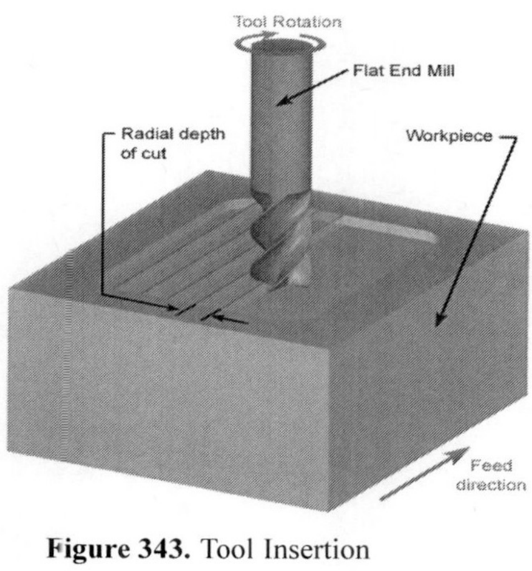


Figure 343. Tool Insertion

## Precaution

1. Always used the sharp edge cutter.
2. Always stay at the machine during running.
3. Don't use a machine at high.
4. Must wear safety gear before starting the operation.
5. You must have at least one person during the operation for any need arises.
6. Check all equipment before any operation.

## Milling Machine Objective Questions

1. In the end mill cutter, flutes of the milling bit are.
   **a) Deep helical grooves running up the cutter**
   b) Sharp Blade along the Edge
   c) Holes
   d) None of the above
2. Following milling machine is used to cut gears.
   **a) Horizontal Milling Machine** b) Vertical Milling Machine
   c) Both a and b d) None of the above
3. In the bed mill, the table moves to the spindle's axis.
   a) Only Parallel
   **b) Only Perpendicular**
   c) Both are Perpendicular and Parallel
   d) Inclined
4. In the vertical milling cutter is held in _______
   **a) The Spindle Rotates on its axis**
   b) Spindle Reciprocates
   c) Arbor rotates on its Axis
   d) Arbor Reciprocates
5. The use of coatings on milling cutters will increase the_________of the tool.
   **a) Surface Hardness** b) Corrosion Resistance
   c) Surface Finish d) Melting Point
6. A turret mill has a_________spindle and ________ the table is moved to the spindle axis to accomplish cutting.
   a) Stationary, Perpendicular
   b) Moving, Parallel
   **c) Stationary, both Perpendicular and Parallel**
   d) Moving, Perpendicular
7. In the end mill cutter, chips are pulled up
   **a) Flute** b) Tooth
   c) Tip d) Shank

8. Upmilling is the process of removing metal by a cutter which is rotated_________direction of the travel of the workpiece.

   a) In the same  **b) Against the**

   c) Can't say anything  d) None of the mentioned

9. Down milling is the process of removing metal by a cutter which is rotated_________ the direction of the travel of the workpiece.

   **a) In the same**  b) Against the

   c) Can't say anything  d) None of the mentioned

10. Narrow slots can be milled in _________ milling process.

   a) Up milling  **b) Down Milling**

   c) Can't say anything  d) None of the mentioned

## Milling Machine Subjective Questions

1 What are the various types of milling machines?

2 What are the various operations performed by the milling machine?

3. Briefly explain the construction and working of

   a) Simple dividing head

   b) Universal dividing head

4. Write and explain procedures for a simple and compound indexing.

## Milling Machine Solved Numericals

1. Calculate feed per tooth, if table feed per minute (feed rate) is 400 mm/min, the number of teeth is 10 and the spindle speed is 500 $min^{-1}$.

   **Solution:**

   We know that, $f = \frac{vf}{z.n}\left(\frac{mm}{m}\right)$

   Where, f is feed per tooth,

   vf is table feed per minute (mm/min), z is number of teeth,

   n is spindle speed ($min^{-1}$)

   $$f = \frac{400}{10 \times 500}(mm) = 0.08\,mm$$

   **Ans:** Feed per tooth is 0.08 mm.

2. Calculate the table feed per minute (feed rate) if, the feed per tooth is 0.4 mm/tooth, the number of teeth is 5 and the spindle speed is 500 $min^{-1}$

   **Solution:**

   We know that, $Vf = fz \ x \ z \ x \ n \ (mm/min)$

   Where, Vf is table feed,

   fz is feed per tooth (mm/tooth).

   z is number of teeth,

   n is spindle speed ($min^{-1}$)

   $$Vf\ 0.4 \times 5 \times 500 \left(\frac{mm}{min}\right) = 1000(mm/min)$$

   **Ans**: Table feed rate is 1000 mm/min.

3. Calculate the cutting speed if the spindle speed is 500 $min^{-1}$ and outer diameter of the cutter is 80 mm.

   **Solution:**

   We know that, $VC = \frac{\pi \times \mathrm{DC} \times \mathrm{n}}{1000}(m/min)$

   Where, VC is cutting speed,

   DC is cutting diameter (mm),

   n is spindle speed ($min^{-1}$)

   $$VC = \frac{\pi \times 80 \times 500}{1000}\left(\frac{m}{min}\right) = 125.66(\mathrm{m/min})$$

   **Ans:** Cutting speed is 125.66 (m/min)

4. Calculate machine time if width is 150 mm, length is 300 mm, feed per tooth is 0.3 mm, number of teeth are 10, spindle speed is 300, cutting speed is 180 m/min, face mill diameter is 200 mm.

   **Solution:**

   We know that,

   total table feed length (L):

   $L$ = *length of material (l) + face mill diameter (DC)*

   $L = 300\ mm + 200\ mm = 500\ mm$

   And, table feed per minute (vf): $Vf = fz \ x \ z \ x \ n \ (mm/min)$

   $$Vf\ 0.3 \times 10 \times 300 \left(\frac{mm}{min}\right) = 900(mm/min)$$

Now, $Tc = \frac{L}{vf}(min)$

$Tc = \frac{500}{900}(min)$ 0.55 (*min.*) = 0.55 x 60 (*sec*) = 33.33 (*sec*)

**Ans:** Machining timing is 33.33 sec.

5. Calculate the table feed per minute (feed rate) if, the feed per tooth is 0.5 mm/tooth, the number of teeth is 10 and the spindle speed is 300 $min^{-1}$.

**Solution:**

We know that, *Vf = fz x z x n (mm/min)*

Where, Vf is table feed,

fz is feed per tooth (mm/tooth).

z is number of teeth,

n is spindle speed ($min^{-1}$)

$$Vf\ 0.5 \times 10 \times 300 \left(\frac{mm}{min}\right) = 1500(\text{mm/min})$$

**Ans:** Table feed rate is 1500 mm/min.

## Milling Machine Unsolved Numericals

1. Calculate feed per tooth, if table feed per minute (feed rate) is 200 mm/min, the number of teeth is 8 and the spindle speed is 250 $min^{-1}$.

   **Ans:** feed per tooth is 0.1 mm.

2. Calculate the table feed per minute (feed rate) if, the feed per tooth is 0.2 mm/tooth, the number of teeth is 8 and the spindle speed is 600 $min^{-1}$.

   **Ans:** Table feed rate is 960 mm/min.

3. Calculate the cutting speed if the spindle speed is 600 and outer diameter of the cutter is 70 mm.

   **Ans:** Cutting speed is 131.94 (m/min).

4. Calculate machine time if width is 150 mm, length is 250 mm, feed per tooth is 0.2 mm, number of teeth are 10, spindle speed is 500 $min^{-1}$, cutting speed is 180 m/min, face mill diameter is 200 mm.

   **Ans:** Machining timing is 27 sec.

5. Calculate machine time if width is 150 mm, length is 200 mm, feed per tooth is 0.5 mm, number of teeth are 8, spindle speed is 200 $min^{-1}$, cutting speed is 180 m/min , face mill diameter is 200 mm.

   **Ans:** Machining timing is 30 sec.

# 11

# Drilling Machine

## Introduction

**Figure 344.** Drilling Machine

In the process of creating a hole in the metal surface through a drill, we use a drilling machine. We use a tapper shank and parallel shank to tie the drill to the drill machine. We can also use a drilling machine to do extra experiments i.e., tapping rimming counter shrinking and counterboring. In simple language, we also say that a, it is a machine which is wildly used over world for making a circular hole in various metal workpiece for making desired design or structure.

## Types of Drilling Machines

There are two types of drilling machines:

1. Portable drilling machine.
2. Stationary drilling machine.

## Portable Drilling Machine

**Figure 345.** Portable Drilling Machine

The machine can be easily delivered from one place to another place. They are called portable drilling machines. These are light-duty machines. Small size drills on these are tied. Heavy jobs that cannot be brought to the machine table, in these cases are used. They do not need to be fixed anywhere.

## Stationary Drilling Machine

These types of machines are fixed permanently above the floor in the workshop. The job is taken to these machines so that drilling etc. can be done. These are heavy-duty machines. We can tie small and big size drills on them. Holes are made on them in pure size. They are driven at different speeds for Drilling.

They also have a coolant system. For drilling, we must put a workpiece on the table.

1. **Bench Type Sensitive Drilling Machine:** This type of machine is driven by electricity. This machine is mounted on the bench that's why we call bench-type sensitive drilling machine. We can use a drill bit up to 12.5 mm. We can tie the drill bit with the help of a drill chuck or taper shank. The table can work in a vertical direction up or down. The spindle can run at various types of speeds. The feed can be given by the hands. This cannot run in automatic mechanisms.

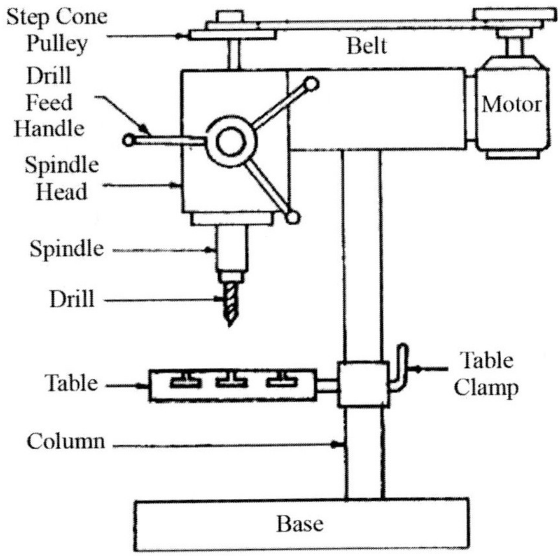


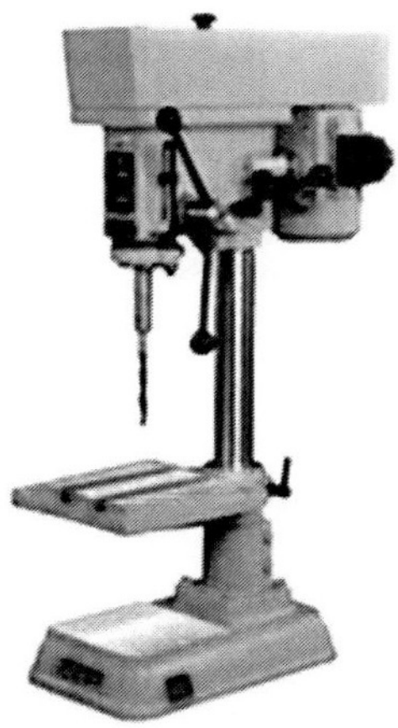

**Figure 346.** Sensitive Drilling Machine

2. **Pillar Type Drill Machine:** This machine work mechanism is the same as the bench type sensitive machine but the difference is that pillar height is bigger than the bench type drill machine. This machine is fixed on the floor. The drill bit can fix up to 25 mm. rack and pinion are used for the worktable to work in an upward and downward direction.

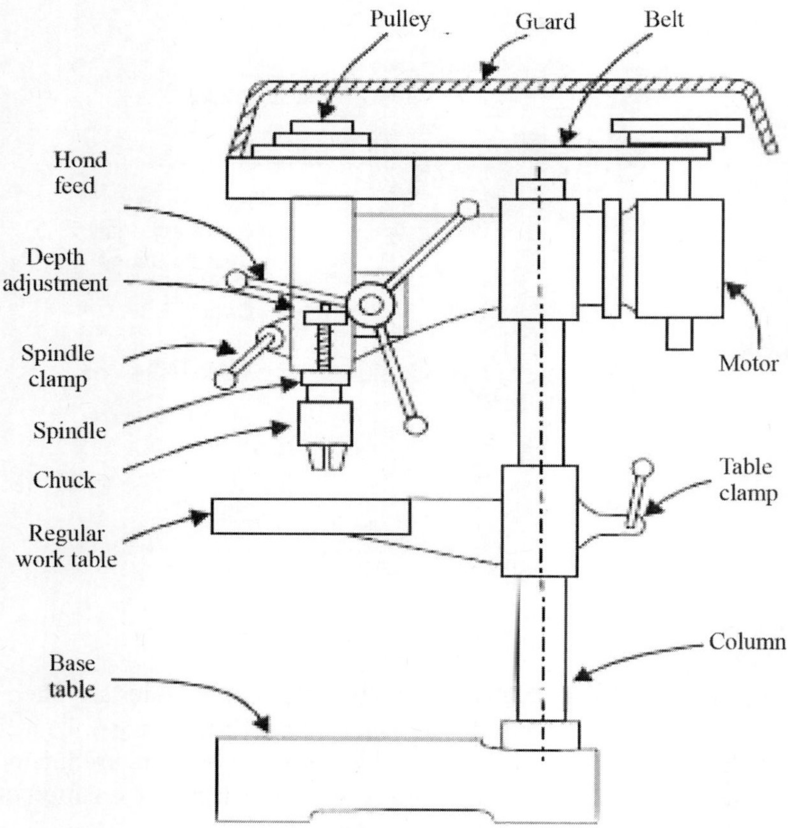


**Figure 347.** Pillar Type Drill Machine

3. **Standard Drill Press:** This machine is used for heavy work. With this machine, small and bigger holes can be made on the metal surface. The different types of operation can be done by the drilling machine i.e., rimming taping because we can use the spindle in reverse. This machine can also run in automatic feed. The column can be square or round.

   The Head is fixed above the column in which the spindle gear and the motor are installed. The working table can be square, round or rectangular. A coolant pump is already installed in this machine.

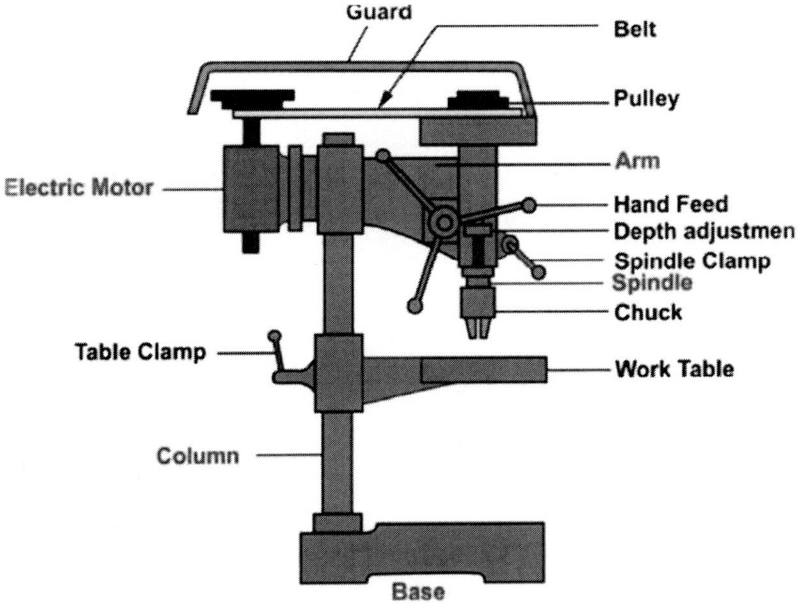


**Figure 348.** Drill Press

4. **Gang Drilling Machine:** In this machine, multiple spindles are fitted they have only one working table. 2 to 8 spindles can be fit in a gang drill machine, every spindle has its motor. We use this machine for more production for sequence- wise operations i.e., drilling, counter shrinking, counterboring, and reaming. Every spindle has its cutting tool.

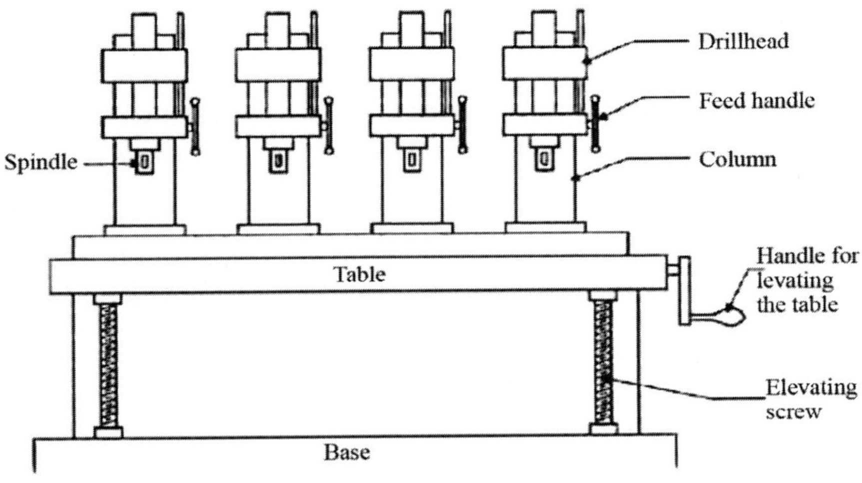


**Figure 349.** Gang Drilling Machine

5. **Multiple Spindle Drilling Machines:** This machine has 4 to 48 spindles. by a single motor. All spindles move. Each spindle is connected by girders. Thus, all the spindles move simultaneously a time. It is used in mass production.

**Figure 350.** Multiple Spindle Drilling Machines

6. **Radial Drilling Machine:** It is a heavy machine. It is used in big factories. By this, drilling is done from small jobs to big jobs. The pillar remains fitted, on which the arm is mounted to the arm can be set anywhere up and down. There is a carrier on the arm, which is used to cover the entire arm. Drilling can be done anywhere by sliding back and forth. It has a gear system.

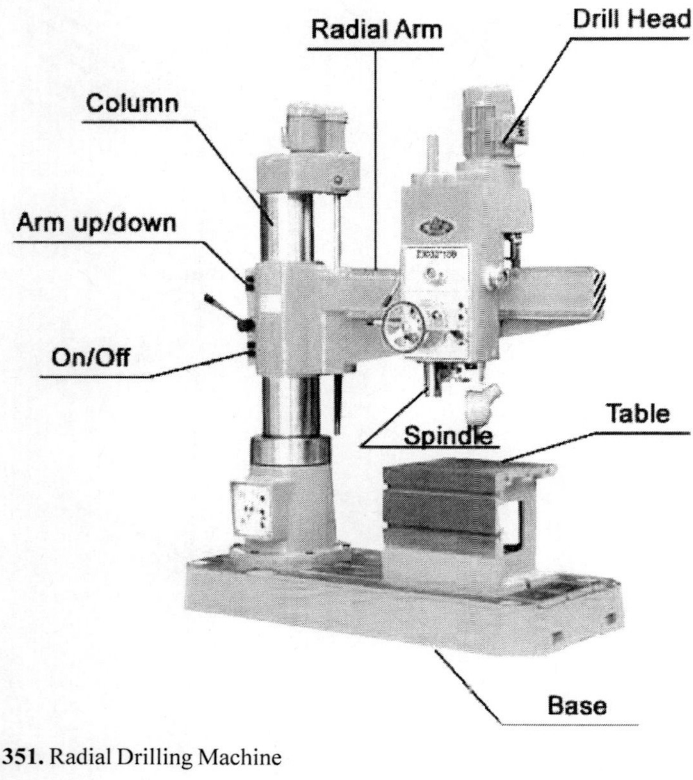


**Figure 351.** Radial Drilling Machine

## Specifications of Drill Machine

If the drill machine is portable then:

1. Hand Drill Press 6 mm.
2. Ratchet Brace 12 mm.
3. Portable Electric Drill Press 220V 12 mm Size.

## Types of Drill Bits

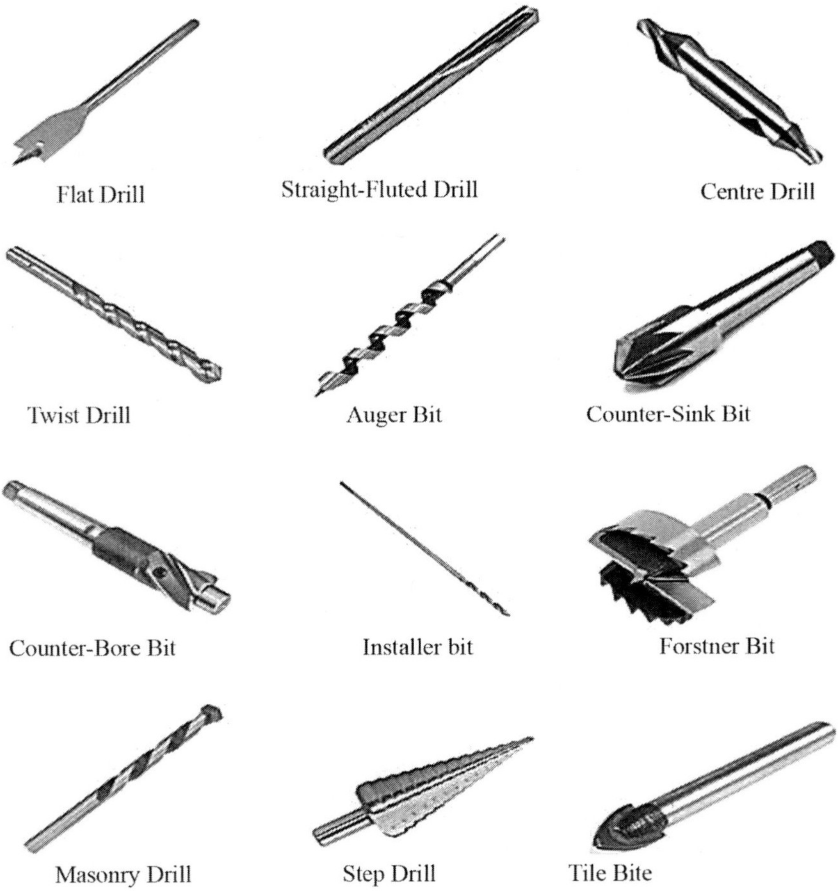


**Figure 352.** Types of Drill Bits

If the drill machine is stationary type, its description is to be done as follows:

1. Standard taper number.
2. Morse taper 2-3 number.
3. Belt Drive or Gear Drive.
4. The maximum vertical distance between spindle and table.
5. Size and shape of the table.
6. Distance from center of spindle to the pillar.
7. Speed of spindle and RPM.
8. Horsepower of the motor.
9. Arm length and pillar diameter in radial drilling machine.
10. The total weight of the machine and the space it occupies.

## Working Principles of A Drilling Machine

The drilling machine is working on the principle that when a power is given to a motor then the motor starts rotating which is connected to the spindle at the head of the machine, than spindle starts rotating and thereby the cutting tool rotates and the various operation takes place, for performing this many types of machines are used in this. And this is work on rack and pinion mechanisms mainly.

## Classification of Drilling Machine Parts

1. **Base:** This is an important part of the machine as the whole system of the machine is concentrated over this, and all the workload or weight is upon the base.

2. **Column:** This part of drilling is used for supporting the head of a drilling machine. The column of a machine is mainly made of hard material in a circular shape.

3. **Spindle:** The work of this part is to revolve the cutting tool for performing a drilling operation.

4. **Sleeve:** The work of the sleeve is that it does not revolve but slides in its own axis of rotation the sleeve carries a spindle, which carries a cutting tool.

5. **Head:** The head it is the upper part of a machine which mainly consists of a spindle, sleeve, electric motor and feed mechanism

## Precautions While Working on A Drilling Machine

The following precautions should be taken while drilling:

1. Drilling Cutting Angle and Lip Clearance: The drill should be ground according to the metal being used.
2. Both the lip of the drill should be of equal length.
3. Well placed on the table of the job machine and in the spindle of the drill machine should be tied.
4. The speed of the machine should be kept according to the size of the drill.
5. The feed should be given according to the hardness of the metal.
6. Equal chips should come from both the flutes while drilling. The drill should be re-grinded.

## Factors for Determining The Speed of Drill Machine

1. **The hardness of the metal:** If the metal is harder the drill speed is kept low while for soft metals the speed is more are kept.
2. **Diameter of the drill:** If the diameter of the drill is large, then the speed is less for it and if the diameter is smaller than the speed will be kept more.
3. **Type and condition of drill machine:** If the drilling machine is portable and in good condition. If not, then the speed is less and if the machine is stationary and the condition is also good then the speed more can be kept.
4. **Hole depth:** The speed is kept low for low and the speed is kept high for less depth.
5. **Arrangement of coolant:** If there is a coolant system, then the speed should be high and if not, then the speed should be kept low.
6. **Arrangement of Clamping:** If the machine is properly clamped on the table, then the speed is high and if there is no proper clamp, then the speed is kept low.

## Experiment 1

**Aim:** To study the drilling operation of drilling machines.

**Materials Required:** Wood, Metal surface (work piece) of mild steel or castiron.

### Tools Required

- Drilling machine.
- Drill bit of high-speed steel or high carbon steel.
- Sleeve.
- Hammer.

**Theory:** In this type of operation drill bit throughout through metal surface.

### Procedure

1. First we have to fit the drill bit in the drilling machine spindle.
2. Then we have to tight the work piece in the table vice.
3. With the help of drawing drilling operations can be done.
4. Give coolant according to the metal we use.

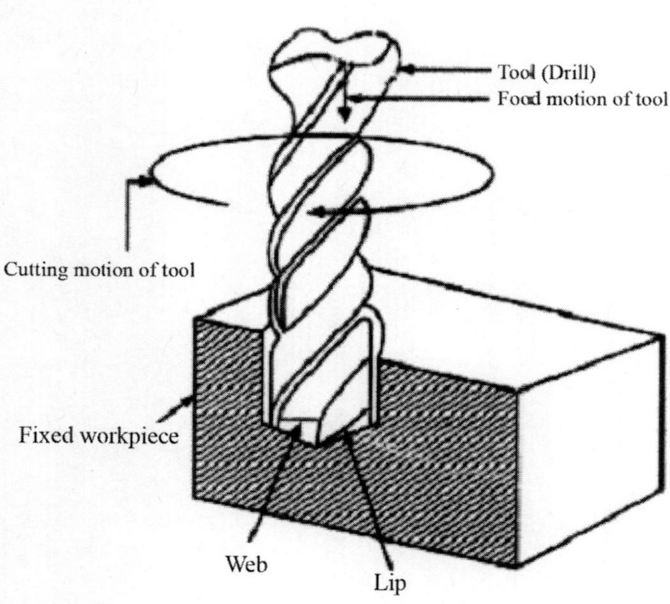


**Figure 353.** Drilling Operation of Drilling Machine

**Result:** Drill hole is created with the help of drill bit on work piece.

## Precautions

1. Pilot drill before drilling a larger size. That is, a small sage should be drilled.
2. Deploy the appropriate amount of coolant according to the metal while drilling needed.
3. Repeatedly ejecting the drill while drilling at greater depths. It should be taken so that the chips can come out.
4. While cutting holes in thin sheets, the wooden packing should be kept at the bottom.
5. When the metal comes to be drilled, the feed should be reduced.

## Experiment 2

**Aim:** To study tapping operation of drilling machine.

**Materials Required:** Wood, metal surface (workpiece) of mild steel or castiron.

**Tools Required:** Drilling machine, Drill tap, sleeve & hammer.

**Theory:** In this action, the drill hole is drilled through the tap inside the bangles are cut. Machine tap is used on drilling machines.

### Procedure

1. First we have to fit the tap tool in the drilling machine spindle.
2. Then we have to tight the work piece in the table vice.
3. With the help of drawing tapping operation can be done.
4. Give lubricant according to the metal we use.

   **Result:** Internal thread is made out of tapping tool.

**Figure 354.** Tapping Operation of Drilling Machine

### Precautions

1. Pilot drill before drilling a larger size. That is, a small sage should be drilled.
2. Deploy the appropriate amount of coolant according to the metal while drilling needed.
3. Repeatedly ejecting the drill while drilling at greater depths. It should be taken so that the chips can come out.
4. While cutting holes in thin sheets, the wooden packing should be kept at the bottom.
5. When the metal comes to be drilled, the feed should be reduced.

## Experiment 3

**Aim:** To study reaming operation of the drilling machine.

**Materials Required:** Wood, Metal surface (work piece) of mild steel or castiron.

### Tools Required

- Drilling machine
- Reamer
- Sleeve
- Hammer

**Theory:** Drill hole corrected by machine reamer. It is made in high finishing.

### Procedure

1. First we have to fit a reamer in the drilling machine spindle.
2. Then we have to tight the workpiece in the table vice.
3. With the help of drawing reaming operation can be done.
4. Give coolant according to the metal we use.

   **Result:** More accurate drill hole is made. And this can also use for more accurate shaft.

### Precautions

1. Pilot drill before drilling a larger size. That is, a small sage should be drilled.
2. Deploy the appropriate amount of coolant according to the metal while drilling needed.
3. Repeatedly ejecting the drill while drilling at greater depths. It should be taken so that the chips can come out.
4. While cutting holes in thin sheets, the wooden packing should be kept at the bottom.

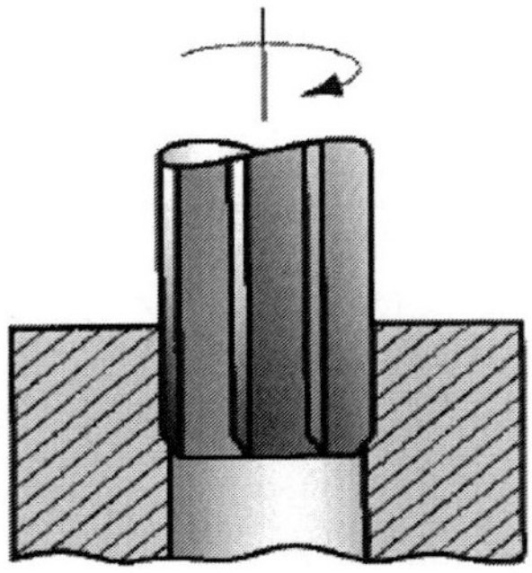

**Figure 355.** Tip of Drill

5. When the metal comes to be drilled, the feed should be reduced.

### Experiment 4

**Aim:** To study Counter Boring operation of the drilling machine.

**Materials Required:** Wood, metal surface (work piece) of mild steel or castiron.

### Tools Required

- Drilling machine
- Counter boring tool
- Sleeve
- Hammer

**Theory:** The diameter of the end is enlarged so that the job surface is flat be able to fit Its surface has to be leveled. A drill of larger size than the hole is used, the diameter of which is must be equal to or something greater than the diameter.

### Procedure

1. First we have to fit the boring tool in the drilling machine spindle.
2. Then we have to tight the work piece in the table vice.
3. With the help of a drawing counter boring operation can be done.
4. Give coolant according to the metal we use.
5. It is the process of enlarging the hole in the work piece.

   **Result:** The end of hole is enlarged cylindrically.

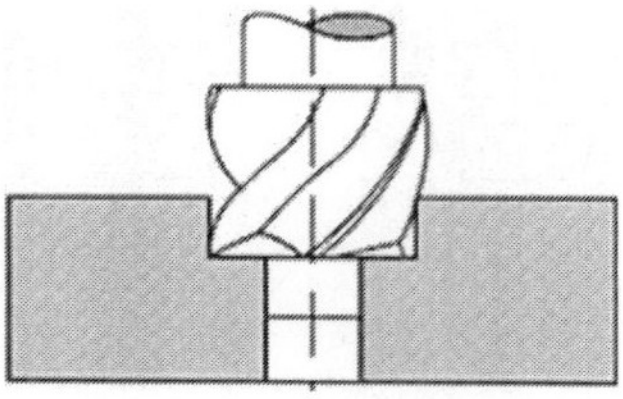

**Figure 356.** Enlarging of Hole

### Precautions

1. Pilot drill before drilling a larger size that is, a small size should be drilled.
2. Deploy the appropriate amount of coolant according to the metal while drilling needed.
3. Repeatedly ejecting the drill while drilling at greater depths. It should be taken so that the chips can come out.
4. While cutting holes in thin sheets, the wooden packing should be kept at the bottom.
5. When the metal comes to be drilled, the feed should be reduced.

## Experiment 5

**Aim:** To study the Spot Facing operation of the drilling machine.

**Materials Required:** Wood, metal surface of mild steel or cast iron.

### Tools Required

- Drilling machine
- Spot facing tool
- Sleeve
- Hammer

**Theory:** In this action, the top surface of the hole is cleaned and made flat at an angle of 90" with the axle. This action allows the nuts or washers to fit snugly against the surface of the job.

### Procedure

1. First we have to fit spot facing tool in the drilling machine spindle.
2. Then we have to tight the work piece in the table vice.
3. With the help of drawing spot facing operation can be done
4. It is used for providing work piece a smooth surface

   **Result:** The top surface of work piece is finished.

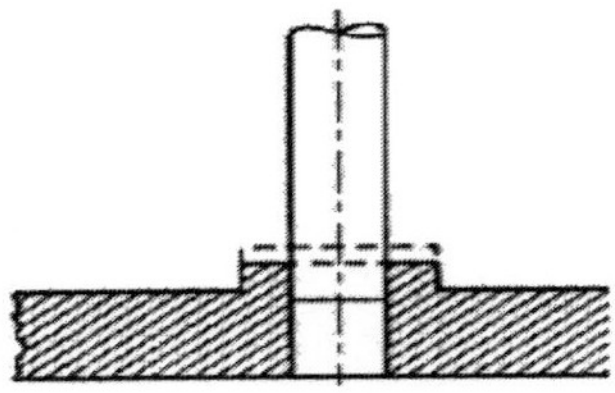

**Figure 357.** Top Surface of Drilling Machine

### Precautions

1. Pilot drill before drilling a larger size. That is, a small sage should be drilled.
2. Deploy the appropriate amount of coolant according to the metal while drilling needed.
3. Repeatedly ejecting the drill while drilling at greater depths. It should be taken so that the chips can come out.
4. While cutting holes in thin sheets, the wooden packing should be kept at the bottom.

## Experiment 6

**Aim:** To study the Boring operation of drilling machines.

**Materials Required:** Wood, metal surface (wcrk piece) of mild steel or castiron.

### Tools Required

- Drilling machine
- Boring tool
- Sleeve
- Hammer

**Theory:** This action is also done on the drilling machine. It is used to make the hole of larger size. This action is done only when drill size is not available.

### Procedure

1. First we have to fit the boring tool in the drilling machine spindle.
2. Then we have to tight the work piece in the table vice.
3. With the help of drawing boring operations can be done.
4. It is used to make holes larger.

   **Result:** The drill hole is enlarged by boring tool.

### Precautions

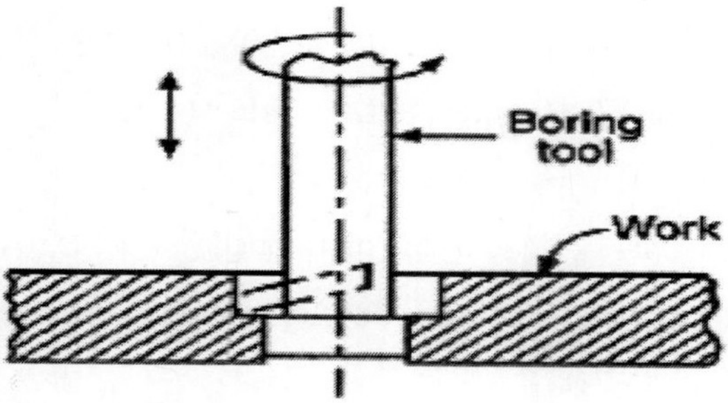


**Figure 358.** Boring Tool

1. Pilot drill before drilling a larger size that is, a small sage should be drilled.
2. Deploy the appropriate amount of coolant according to the metal while drilling needed.
3. Repeatedly ejecting the drill while drilling at greater depths. It should be taken so that the chips can come out.
4. While cutting holes in thin sheets, the wooden packing should be kept at the bottom.
5. When the metal comes to be drilled, the feed should be reduced.

## Drilling Machine Objective Questions

1. Which of the following device is not used for holding the work in a drilling machine?

   a) Step Block  b) Drill Jigs

   c) Both step Block and drill jigs  **d) None of the mentioned**

2. The drilling machine was primarily designed to originate a.

   a) Flat Surface  b) Cylindrical Surface

   c) Taper Surface  **d) None of the mentioned**

3. For drilling operation, the drill rotates with pressure if job is held on earth.

   a) Upward

   **b) Downward**

   c) 45 degree inclined to both planes

   d) None of the mentioned

4. Which of the following method is not used for holding tool?

   a) By a Sleeve  b) By Chucks

   c) By a Socket  **d) None of the mentioned**

5. Floating holder is included in method of tool holding devices.

   a) By directly fitting in the Spindle

   **b) By special Attachments**

   c) By a Socket

   d) None of the mentioned

6. Which of the following operation is performed to provide recess for bolt heads or nuts?

   **a) Counter Boring**  b) Spot Facing

   c) Tapping  d) None of the mentioned

7. Which of the following operation is carried out to make the hole dimensionally more accurate?

   **a) Reaming**  b) Tapping

   c) Boring  d) None of mentioned

8. To produce more accurate holes, which of the following operation should be performed third?

   a) Drilling b) Reaming

   **c) Centering** d) Boring

9. Which of the following operation is performed to ensure the correct location of a hole by making it concentric with the axis of rotation?

   a) Parting b) Tapping

   **c) Boring** d) None of mentioned

10. To produce more accurate holes, which of the following operation should be performed last?

    a) Drilling b) Reaming

    **c) Centering** d) Boring

## Drilling Machine Subjective Questions

Q1 Name different types of drilling machine?

Q2 State various type of work holding devices of drilling machine.

Q3 What are the tool holding devices in drilling machine.

Q4 Differentiate between counter boring and counter sinking operations.

## Drilling Machine Solved Numericals

1. A hole is being drilled with 6 mm drill at a feed of 0.2 mm / revolution and rpm is 300. Calculate the metal removal rate (m) in $mm^3$ / minute.

   **Solution:**

   Here, we have the value of diameter, feed and rpm. So, from the formula, m = (3.14*d*d*f*n) / 4, we can easily get the value of metal removal rate.

   **Ans.** 1695.6mm3/min.

2. The diameter of the drill is 15 mm and rpm of drill is 5. Find the value of cutting speed in meter/minute.

   **Solution:**

   From the formula, cs = (3.14*d*n) /1000.

   **Ans.** 0.628 m/min.

3. Rpm is 50 and feed per revolution (f2) is 10. Then find the value of feed per minute (f1).

   **Solution:**

   Rpm = f1/f2.

   Here rpm = 50 and f2 = 10.

   So value of f1 should be 500.

   **Ans.** f1 = 500.

4. If depth of cut is 40 mm, then find the value of diameter of the drill. Solution:

   Depth of cut = diameter/2.

   **Ans.** Diameter of drill = 80 mm.

5. If t = 5 minute, n = 5 rpm and s = 10 mm. Then find the value of l. t = machining time in minute, l = length of travel of the drill in mm, n = rpm of the drill and s = feed per revolution of the drill in mm.

   **Solution:**

   From the formula

   t = l / (n*s). Here t = 5, n = 5 and s = 10.

   Put all these values with their units in the mentioned equation, we will easily get the answer as 250 mm.

   **Ans.** l = 250 mm.

### Drilling Machine Unsolved Numericals

1. If t = 3 minute, n = 10 rpm and s = 10 mm. Then find the value of t = machining time in minute, l = length of travel of the drill in mm, n = rpm of the drill and s = feed per revolution of the drill in mm.

   **Ans. 300**

2. Feed per minute (f1) is 30 and rpm (n) is 6. Then find the value of feed per revolution (f2).

   **Ans. 5**

3. If cs = 0.846 and d = 45, then find the value of n. d = diameter of drill in mm, n = rpm and cs = cutting speed in mm/minute.

   **Ans. 18.8**

4. If drill is 5 mm, feed is 0.7 mm/revolution and metal removal rate is 80 mm3/ minute. Then find out the value of rpm.

   **Ans. 5.82**

5. The diameter of the drill is 30 mm and rpm of drill is 10. Find the value of cutting speed in meter/minute.

   **Ans. 1.02**

# 12

# Unit and Conversion Factors

## I. Decimal Multiples and Sub-Multiples

| Name | Symbol | Equivalent | | Name | Symbol | Equivalent |
|---|---|---|---|---|---|---|
| tera | T | $10^{12}$ | | deci | d | $10^{-1}$ |
| giga | G | $10^{9}$ | | centi | c | $10^{-2}$ |
| mega | M | $10^{6}$ | | milli | m | $10^{-3}$ |
| kilo | k | $10^{3}$ | | micro | μ | $10^{-6}$ |
| hecto | h | $10^{2}$ | | nano | n | $10^{-9}$ |
| deca | – | 10 | | pico | p | $10^{-12}$ |

## II. Description of Units

### Mechanical Units

| Quantity | cgs | mks |
|---|---|---|
| Length | centimeter | meter |
| Mass | gram | kilogram |
| Time | second | second |
| Force | dyne | newton |
| Work, Energy | erg | joule |
| Power | – | watt |
| Dynamic Viscosity | poise | – |
| Kinematic Viscosity | stoke | – |

### Electric and Magnetic Units

The esu and emu unit systems are cgs systems.

esu denotes "electrostatic unit", sometimes given prefix "stat", e.g. statcoulomb emu denotes "electromagnetic unit", sometimes given prefix "ab", e.g. abcoulomb Some emu units have special names:

| Quantity | emu | mksq |
|---|---|---|
| Magnetic Flux, $\phi$<br>Magnetic Field Strength, B<br>Magnetomotive Force, F Magnetic Vector, H | line maxwellgauss<br>gilbert oersted | weber<br>weber/m$^2$ amp-turn<br>amp-turn/m. |

## mksq Dimensions

| | |
|---|---|
| Length | L |
| Mass | M |
| Time | T |
| Current | Q/T |
| Charge | Q |

## III. Equivalent Units mksq System

| | | |
|---|---|---|
| 1 newton | = | 1 kilogram-meter/(second)$^2$ |
| 1 volt | = | 1 newton-meter/coulomb |
| 1 amp | = | 1 coulomb/second |
| 1 joule | = | 1 newton-meter = 1 coulomb-volt |
| 1 weber | = | 1 volt-second |
| 1 farad | = | 1 coulomb/volt |
| 1 henry | = | 1 weber/amp |
| 1 ohm | = | 1 volt/amp |
| 1 watt | = | 1 joule/sec |

## IV. Dimensions of esu and emu Electric and Magnetic Quantities

The fundamental dimensions in both systems are M, L, T. cgs units used.

| Quantity | Symbol | esu | | | emu | | | emu* |
|---|---|---|---|---|---|---|---|---|
| | | $M^XL^YT^Z$ | | | $M^XL^YT^Z$ | | | esu |
| | | x | y | z | x | y | z | |
| Charge | q | ½ | 3/2 | -1 | ½ | ½ | 0 | c |
| Field Intensity | E | ½ | -½ | -1 | ½ | -3/2 | 0 | 1/c |
| Elec. Displacement | D | ½ | -½ | -1 | ½ | ½ | -2 | c |
| Charge Density | ρ | ½ | -3/2 | -1 | ½ | -5/2 | 0 | c |
| Current Density | j | ½ | -½ | -2 | ½ | -3/2 | -1 | c |
| Elec. Potential | V | ½ | ½ | -1 | ½ | 3/2 | -2 | 1/c |
| Total Current | I | ½ | 3/2 | -2 | ½ | ½ | -1 | c |

| | | | | | | | | |
|---|---|---|---|---|---|---|---|---|
| Mag. Field Strength | B | ½ | -3/2 | 0 | ½ | -½ | -1 | 1/c |
| Mag. Vector | H | ½ | ½ | -2 | ½ | -½ | -1 | c |
| Permittivity | ε | 0 | 0 | 0 | 0 | -2 | 2 | $c^2$ |
| Permeability | μ | 0 | -2 | 2 | 0 | 0 | 0 | $1/c^2$ |
| Conductivity | σ | 0 | 0 | -1 | 0 | -2 | 1 | $c^2$ |
| Capacitance | C | 0 | 1 | 0 | 0 | -1 | 2 | $c^2$ |
| Inductance | L | 0 | -1 | 2 | 0 | 1 | 0 | $1/c^2$ |
| Resistance | R | 0 | -1 | 1 | 0 | 1 | -1 | $1/c^2$ |

*c = velocity of light (free space) in cm/sec ≈ $3 \times 10^{10}$

Thus: 1 emu of charge = $2.998 \times 10^{10}$ esu of charge
or 1 abcoulomb = $2.998 \times 10^{10}$ statcoulomb

## V. Dimensions snd Units for Physical Quantities mksq System

### A. Mechanical Quantities

| **Quantity** | **Dimensions** | **Derived Units** |
|---|---|---|
| Acceleration | $LT^{-2}$ | meter/sec$^2$ |
| Angle | 0 | radian |
| Angular Accleration | $T^{-2}$ | radian/sec$^2$ |
| Angular Momentum | $ML^2T^{-1}$ | kgm-meter$^2$/sec |
| Angular Velocity | $T^{-1}$ | radian/sec |
| Area | $L^2$ | meter$^2$ |
| Energy | $ML^2T^{-2}$ | joule |
| Force | $MLT^{-2}$ | newton |
| Frequency | $T^{-1}$ | cycle/sec |
| Gravitational Field Strength | $LT^{-2}$ | newton/kgm |
| Length | L | meter |
| Mass | M | kilogram |
| Mass Density | $ML^{-3}$ | kgm/meter$^2$ |
| Momentum | $MLT^{-1}$ | kgm-meter/sec |
| Power | $ML^2T^{-3}$ | watt |
| Pressure | $ML^{-1}T^{-2}$ | newton/meter$^2$ |
| Time | T | second |
| Torque | $ML^2T^{-2}$ | newton/meter |

| | | |
|---|---|---|
| Velocity | $LT^{-1}$ | meter/sec |
| Viscosity (Dynamic) | $ML^{-1}T^{-1}$ | kgm/meter-sec |
| Viscosity (Kinematic) | $L^2T^{-1}$ | meter$^2$/sec |
| Volume | $L^3$ | meter$^3$ |
| Wave Length | L | meter |
| Work | $ML^2T^{-2}$ | joule |

## B. Thermal Quantities*

| Quantity | Dimensions | Derived Units |
|---|---|---|
| Enthalpy | $ML^2T^{-2}$ | joule |
| Entropy | $ML^2T^{-2}\theta^{-1}$ | joule/K° |
| Gas Constant | $L^2T^{-2}\theta^{-1}$ | joule/kgm- K° |
| Internal Energy | $ML^2T^{-2}$ | joule |
| Specific Heat | $L^2T^{-2}\theta^{-1}$ | joule/kgm-K° |
| Temperature | $\theta$ | K° |
| Thermal Conductivity | $MLT^{-}3\theta^{-1}$ | watt/meter- K° |
| Thermal Diffusivity | $L^2T^{-1}$ | meter$^2$/sec |
| Heat Transfer Coefficient | $MT^{-}3\theta^{-1}$ | watt/meter$^2$- K° |

*The dimension of temperature is θ; the unit is K°

## C. Electric and Magnetic Quantities

| Quantity | Symbol | Dimensions | Derived Units |
|---|---|---|---|
| Charge | q | Q | coulomb |
| Field Intensity | E | $MLT^{-2}Q^{-1}$ | volt/meter |
| Elec. Displacement | D | $L^{-2}Q$ | coulomb/meter$^2$ |
| Charge Density | ρ | $L^{-3}Q$ | coulomb/meter$^2$ |
| Current Density | j | $L^{-2}T^{-1}Q$ | amp/meter$^2$ |
| Elec. Potential | V | $ML^2T^{-2}Q^{-1}$ | volt |
| Total Current | I | $T^{-1}Q$ | coulomb/sec |
| Mag. Field Strength | B | $MT^{-1}Q^{-1}$ | weber/meter$^2$ |
| Mag. Vector | H | $L^{-1}T^{-1}Q$ | amp(turn)/meter |
| Permittivity | ε | $M^{-1}L^{-3}T^2Q^2$ | farad/meter |
| Permeability | μ | $MLQ^{-2}$ | henry/meter |
| Conductivity | σ | $M^{-1}L^{-3}TQ^2$ | 1/ohm-meter |
| Capacitance | C | $M^{-1}L^{-2}T^2Q^2$ | farad |
| Inductance | L | $ML^2Q^{-2}$ | henry |
| Resistance | R | $ML^2T^{-1}Q^{-2}$ | ohm |

## VI. Conversion of mksq Units to Gaussian Units

| Quantity | mksq Unit | | Conversion Factor × Gaussian Unit* | | |
|---|---|---|---|---|---|
| q | coulomb | = | $10^{-1}$ c | statcoulomb | (esu) |
| E | volt/meter | = | $10^{6}$/c | statvolt/cm | (esu) |
| D | coulomb/meter$^2$ | = | $4\pi \times 10^{-5}$ c | lines/cm$^2$ | (esu) |
| ρ | coulomb/meter$^3$ | = | $10^{-7}$ c | statcoulomb/cm$^3$ | (esu) |
| j | amp/meter$^2$ | = | $10^{-5}$ | abamp/cm$^2$ | (emu) |
| V | volt | = | $10^{8}$/c | statvolt | (esu) |
| I | coulomb/sec : = amp | = | $10^{-1}$ | abamp | (emu) |
| B | weber/meter$^2$ | = | $10^{4}$ | gauss | (emu) |
| H | amp-turn/meter | = | $4\pi \times 10^{-3}$ | oersted | (emu) |
| μ | farad/meter | = | $4\pi \times 10^{-11}c^2$ | — | (esu) |
| ε | henry/meter | = | $10^{7}/4\pi$ | — | (emu) |
| σ | 1/ohm-meter | = | $10^{-11}$ | 1/abohm-cm | (emu) |
| C | farad | = | $10^{-9}c^2$ | statfarad | (esu) |
| L | henry | = | $10^{9}$ | abhenry | (emu) |
| R | ohm | = | $10^{9}$ | abohm | (emu) |

*c = vel. of light (free space) in cm/sec ≈ $3 \times 10^{10}$

### Use of table:

1 coulomb = $10^{-1}$ ($3 \times 10^{10}$) statcoulomb = $3 \times 10^{9}$ statcoulomb

## VI. Conversion Factors

**Note :** mksq Units are Capitalized

Use of Tables: Example

1 degree = $2.778 \times 10^{-3}$ revolutions

so, 16.7° = $16.7 \times 2.778 \times 10^{-3}$ revolutions

### A. Plane Angle

| | o | ′ | ″ | RADIAN | rev |
|---|---|---|---|---|---|
| 1 degree = | 1 | 60 | 3600 | $1.745 \times 10^{-2}$ | $2.778 \times 10^{-3}$ |
| 1 minute = | $1.667 \times 10^{-2}$ | 1 | 60 | $2.909 \times 10^{-4}$ | $4.630 \times 10^{-5}$ |
| 1 second = | $2.778 \times 10^{-4}$ | $1.667 \times 10^{-2}$ | 1 | $4.848 \times 10^{-4}$ | $7.716 \times 10^{-7}$ |
| 1 RADIAN = | 57.30 | 3438 | $2.063 \times 10^{5}$ | 1 | 0.1592 |
| 1 revolution = | 360 | $2.16 \times 10^{4}$ | $1.296 \times 10^{5}$ | 6.283 | 1 |

1 revolution = 2 π RADIANS = 360°, 1° = 60′ = 3600″

## B. Solid Angle

1 sphere = 4 $\pi$ steradians = 12.57 steradians

## C. Length

| | cm | METER | km | in | ft | mile |
|---|---|---|---|---|---|---|
| 1 centimeter = | 1 | 10 | $10^{-5}$ | 0.3937 | $3.281 \times 10^{-2}$ | $6.214 \times 10^{-6}$ |
| 1 METER = | 100 | 1 | $10^{-3}$ | 39.37 | 3.281 | $6.214\times 10^{-4}$ |
| 1 kilometer = | $10^5$ | 1000 | 1 | $3.937 \times 10^{-4}$ | 3281 | 0.6214 |
| 1 inch = | 2.540 | $2.540 \times 10^{-2}$ | $2.540 \times 10^{-3}$ | 1 | $8.333 \times 10^{-2}$ | $1.578 \times 10^{-5}$ |
| 1 foot = | 30.48 | 0.3048 | $3.048 \times 10^{-4}$ | 12 | 1 | $1.894 \times 10^{-4}$ |
| 1 statute mile = | $1.609 \times 10^5$ | 1609 | 1.609 | $6.336 \times 10^4$ | 5280 | 1 |

1 foot = 1200/3937 meter

1 meter = 3937/1200 ft

1 angstrom (Å) = $10^{-10}$ meter

1 X-unit = $10^{-13}$ meter

1 micron = $10^{-6}$ meter

1 millimicron (mu) = $10^{-9}$ meter

1 light-year = $9.460 \times 10^{12}$ km

1 par-sec = $3.084 \times 10^{13}$ km

1 fathom = 6 ft

1 yard = 3 ft

1 rod = 16.5 ft

1 mil = $10^{-3}$ in

1 nautical mile = 1852 meters = 1.1508 statute miles

1 nautical mile = 6076.10 ft

## D. Area

| | METER² | cm² | ft² | in² | circ mil |
|---|---|---|---|---|---|
| 1 SQUARE METER = | 1 | $10^4$ | 10.76 | 1550 | $1.974 \times 10^9$ |
| 1 square cm = | $10^{-4}$ | 1 | $1.076 \times 10^{-3}$ | 0.1550 | $1.974 \times 10^5$ |
| 1 square foot = | $9.290 \times 10^{-2}$ | 929.0 | 1 | 144 | $1.833 \times 10^8$ |
| 1 square inch = | $6.452 \times 10^{-4}$ | 6.452 | $6.944 \times 10^{-3}$ | 1 | $1.273 \times 10^6$ |
| 1 circular mil = | $5.067 \times 10^{-10}$ | $5.067 \times 10^{-6}$ | $5.454 \times 10^{-3}$ | $7.854 \times 10^{-7}$ | 1 |

1 square mile = 27,878

1 acre = 43,560 ft$^2$

1 hectare = 2.417 acres

400 ft$^2$ = 640 acres

1 barn = $10^{-28}$ meter$^2$

## E. Volume

| | | METER³ | cm³ | liter | ft³ | in³ |
|---|---|---|---|---|---|---|
| 1 CUBIC METER | = | 1 | $10^6$ | 1000 | 35.31 | $6.102 \times 10^4$ |
| 1 cubic cm | = | $10^{-6}$ | 1 | $1.000 \times 10^{-3}$ | $3.531 \times 10^{-8}$ | $6.102 \times 10^{-2}$ |
| 1 liter | = | $1.000 \times 10^{-3}$ | 1000 | 1 | $3.531 \times 10^{-2}$ | 61.02 |
| 1 cubic foot | = | $2.832 \times 10^{-2}$ | $2.832 \times 10^4$ | 28.32 | 1 | 1728 |
| 1 cubic inch | = | $1.639 \times 10^{-5}$ | 16.39 | $1.639 \times 10^{-2}$ | $5.787 \times 10^{-4}$ | 1 |

1 U.S. fluid gallon = 4 U.S. fluid quarts = 8 U.S. fluid pints

=128 U.S. fluid ounces = 231 in³

1 British imperial gallon = 277.42 in³ (volume of 10 lb $H_2O$ at 62° F)

1 liter = 1000.028 cm³ (volume of 1 kgm $H_2O$ at its maximum density)

## F. Mass

| | | gm | KGM | slug | amu | oz | lb | ton |
|---|---|---|---|---|---|---|---|---|
| 1 gram | = | 1 | 0.001 | $6.852 \times 10^{-5}$ | $6.024 \times 10^{23}$ | $3.527 \times 10^{-2}$ | $2.205 \times 10^{-3}$ | $1.102 \times 10^{-6}$ |
| 1 KILOGRAM | = | 1000 | 1 | $6.852 \times 10^{-2}$ | $6.024 \times 10^{26}$ | 35.27 | 2.205 | $1.102 \times 10^{-3}$ |
| 1 slug | = | $1.459 \times 10^{-4}$ | 14.59 | 1 | $8.789 \times 10^{27}$ | 514.8 | 32.17 | $1.609 \times 10^{-2}$ |
| 1 amu | = | $1.660 \times 10^{-24}$ | $1.660 \times 10^{-27}$ | $1.137 \times 10^{-28}$ | 1 | $5.855 \times 10^{-26}$ | $3.660 \times 10^{-27}$ | $1.829 \times 10^{-30}$ |
| 1 ounce (avoirdupois) | = | 28.35 | $2.835 \times 10^{-2}$ | $1.943 \times 10^{-3}$ | $1.708 \times 10^{25}$ | 1 | $6.250 \times 10^{-2}$ | $3.125 \times 10^{-5}$ |
| 1 pound (avoirdupois) | = | 453.6 | 0.4536 | $3.108 \times 10^{-2}$ | $2.732 \times 10^{26}$ | 16 | 1 | 0.0005 |
| 1 ton | = | $9.072 \times 10^{-5}$ | 907.2 | 62.16 | $5.465 \times 10^{29}$ | $3.200 \times 10^4$ | 2000 | 1 |

**Note** : Portion of table enclosed in heavy lines must be used with caution because those units are not properly mass units but weight equivalents which depend on standard terrestrial acceleration due to gravity, i.e. g.

## G. Density

| | | slug/ft$^3$ | KGM/M$^3$ | gm/cm$^3$ | lb/ft$^3$ | lb/in$^3$ |
|---|---|---|---|---|---|---|
| 1 slug per ft$^3$ | = | 1 | 515.4 | 0.5154 | 32.17 | $1.862 \times 10^{-2}$ |
| 1 KILOGRAM per METER$^3$ | = | $1.940 \times 10^{-3}$ | 1 | 0.001 | $6.243 \times 10^{-2}$ | $3.613 \times 10^{-5}$ |
| 1 gm per cm$^3$ | = | 1.940 | 1000 | 1 | 62.43 | $3.613 \times 10^{-2}$ |
| 1 pound per ft$^3$ | = | $3.108 \times 10^{-2}$ | 16.02 | $1.602 \times 10^{-2}$ | 1 | $5.787 \times 10^{-4}$ |
| 1 pound per in$^3$ | = | 53.71 | $2.768 \times 10^4$ | 27.68 | 1728 | 1 |

**Note** : Portion of table enclosed in heavy lines must be used with caution because those units are not mass-density units but weight-density units which depend on g.

## H. Time

| | yr | day | hr | min | SECOND |
|---|---|---|---|---|---|
| 1 year = | 1 | 365.2 | $8.766 \times 10^3$ | $5.259 \times 10^3$ | $3.156 \times 10^7$ |
| 1 day = | $2.738 \times 10^{-3}$ | 1 | 24 | 1440 | $8.640 \times 10^4$ |
| 1 hour = | $1.141 \times 10^{-4}$ | $4.167 \times 10^{-2}$ | 1 | 60 | 3600 |
| 1 minute = | $1.901 \times 10^{-6}$ | $6.944 \times 10^{-4}$ | $1.667 \times 10^{-2}$ | 1 | 60 |
| 1 SECOND = | $3.169 \times 10^{-8}$ | $1.157 \times 10^{-5}$ | $2.778 \times 10^{-4}$ | $1.667 \times 10^{-2}$ | 1 |

1 year = 365.24219879 days

## I. Speed

| | ft/sec | km/hr | METER/SEC | miles/hr | cm/sec | knot |
|---|---|---|---|---|---|---|
| 1 foot per second = | 1 | 1.097 | 0.3408 | 0.6818 | 30.48 | 0.5925 |
| 1 kilometer per hour = | 0.9113 | 1 | 0.2778 | 0.6214 | 27.78 | 0.5400 |
| 1 METER per SECOND = | 3.281 | 3.600 | 1 | 2.237 | 100 | 1.944 |
| 1 mile per hour = | 1.467 | 1.609 | 0.4770 | 1 | 44.70 | 0.8689 |
| 1 centimeter per sec = | $3.281 \times 10^{-2}$ | $3.600 \times 10^{-2}$ | 0.0100 | $2.237 \times 10^{-2}$ | 1 | $1.944 \times 10^{-2}$ |
| 1 knot = | 1.688 | 1.852 | 0.5144 | 1.151 | 51.44 | 1 |

1 knot = 1 nautical mile/hr

1 mile/min = 88 ft/sec
= 60 miles/hr

## J. Force

| | | dyne | NT | lb | pdl | gf | kgf |
|---|---|---|---|---|---|---|---|
| 1 dyne | = | 1 | $10^{-5}$ | $2.248 \times 10^{-6}$ | $7.233 \times 10^{-5}$ | $1.020 \times 10^{-3}$ | $1.020 \times 10^{-6}$ |
| 1 NEWTON | = | $10^5$ | 1 | 0.2248 | 7.233 | 102.0 | 0.1020 |
| 1 pound | = | $4.480 \times 10^5$ | 4.448 | 1 | 32.17 | 453.6 | 0.4536 |
| 1 poundal | = | $1.383 \times 10^4$ | 0.1383 | $3.108 \times 10^{-2}$ | 1 | 14.10 | $1.410 \times 10^{-2}$ |
| 1 gram-force | = | 980.7 | $9.807 \times 10^{-3}$ | $2.205 \times 10^{-3}$ | $7.093 \times 10^{-2}$ | 1 | 0.001 |
| 1 kilogram-force | = | $9.807 \times 10^5$ | 9.807 | 2.205 | 70.93 | 1000 | 1 |

**Note** : Portion of table enclosed in heavy lines must be used with caution because those units are not force units but weight equivalents of mass which depend on g.

1 kgf = 9.80665 newton 1 lb = 32.17398 poundal

## K. Pressure

| | | atm | dyne/ cm² | inch of water | cm Hg | NT/ METER² | lb/in² | lb/ft² |
|---|---|---|---|---|---|---|---|---|
| 1 atmosphere | = | 1 | $1.013 \times 10^6$ | 406.8 | 76 | $1.013 \times 10^5$ | 14.70 | 2116 |
| 1 dyne per cm² | = | $9.869 \times 10^{-7}$ | 1 | $4.015 \times 10^{-4}$ | $7.501 \times 10^{-5}$ | 0.100 | $1.450 \times 10^{-5}$ | $2.089 \times 10^{-3}$ |
| 1 inch of water at 4°C[a] | = | $2.458 \times 10^{-3}$ | 2.491 | 1 | 0.1868 | 249.1 | $3.613 \times 10^{-2}$ | 5.202 |
| 1 centimeter of mercury at 0°C[a] | = | $1.316 \times 10^{-2}$ | $1.333 \times 10^4$ | 5.353 | 1 | 1333 | 0.1934 | 27.85 |
| 1 NEWTON per METER² | = | $9.869 \times 10^{-6}$ | 10 | $4.015 \times 10^{-3}$ | $7.501 \times 10^{-4}$ | 1 | $1.450 \times 10^{-4}$ | $2.089 \times 10^{-2}$ |
| 1 pound per in² | = | $6.805 \times 10^{-2}$ | $6.895 \times 10^4$ | 27.68 | 5.171 | $6.895 \times 10^3$ | 1 | 144 |
| 1 pound per ft² | = | $4.725 \times 10^{-4}$ | 478.8 | 0.1922 | $3.591 \times 10^{-2}$ | 47.88 | $6.944 \times 10^{-3}$ | 1 |

[a] Where the acceleration of gravity has the standard value 9.80665 meter/sec² 1 bar = $10^6$ dyne/cm² 1 millibar = $10^3$ dyne/cm²

1 torr (mm Hg at 0°C) = $1.93367 \times 10^{-2}$ lb/ft²

## L. Energy, Work, Heat

| | Btu | erg | ft-lb | hp-hr | JOULES | cal | kw-hr | ev | Mev | kgm | amu |
|---|---|---|---|---|---|---|---|---|---|---|---|
| 1 British thermal unit = | 1 | $1.055 \times 10^{10}$ | 777.9 | $3.929 \times 10^{-4}$ | 1055 | 252.0 | $2.930 \times 10^{-4}$ | $6.585 \times 10^{21}$ | $6.585 \times 10^{15}$ | $1.174 \times 10^{-14}$ | $7.074 \times 10^{12}$ |
| 1 erg = | $9.481 \times 10^{-11}$ | 1 | $7.376 \times 10^{-8}$ | $3.725 \times 10^{-14}$ | $10^{-7}$ | $2.389 \times 10^{-8}$ | $2.778 \times 10^{-14}$ | $6.242 \times 10^{11}$ | $6.242 \times 10^{5}$ | $1.113 \times 10^{-24}$ | 670.5 |
| 1 foot-pound = | $1.285 \times 10^{-3}$ | $1.356 \times 10^{7}$ | 1 | $5.051 \times 10^{-7}$ | 1.356 | 0.3239 | $3.766 \times 10^{-7}$ | $8.464 \times 10^{18}$ | $8.464 \times 10^{12}$ | $1.509 \times 10^{-17}$ | $9.092 \times 10^{9}$ |
| 1 horsepower-hour = | 2545 | $2.685 \times 10^{-13}$ | $1.980 \times 10^{6}$ | 1 | $2.685 \times 10^{6}$ | $6.414 \times 10^{5}$ | 0.7457 | $1.676 \times 10^{25}$ | $1.676 \times 10^{19}$ | $2.988 \times 10^{-11}$ | $1.800 \times 10^{16}$ |
| 1 JOULE = | $9.481 \times 10^{-4}$ | $10^{7}$ | 0.7376 | $3.725 \times 10^{-7}$ | 1 | 0.2389 | $2.778 \times 10^{-7}$ | $6.242 \times 10^{18}$ | $6.242 \times 10^{12}$ | $1.113 \times 10^{-17}$ | $6.705 \times 10^{9}$ |
| 1 calorie = | $3.968 \times 10^{-3}$ | $4.186 \times 10^{7}$ | 3.087 | $1.559 \times 10^{-6}$ | 4.186 | 1 | $1.163 \times 10^{-6}$ | $2.613 \times 10^{19}$ | $2.613 \times 10^{13}$ | $4.659 \times 10^{-17}$ | $2.807 \times 10^{10}$ |
| 1 kilowatt-hour = | 3413 | $3.6 \times 10^{13}$ | $2.655 \times 10^{6}$ | 1.341 | $3.6 \times 10^{6}$ | $8.601 \times 10^{5}$ | 1 | $2.247 \times 10^{25}$ | $2.247 \times 10^{19}$ | $4.007 \times 10^{-11}$ | $2.414 \times 10^{16}$ |
| 1 electron volt = | $1.519 \times 10^{-22}$ | $1.602 \times 10^{-12}$ | $1.182 \times 10^{-19}$ | $5.967 \times 10^{-26}$ | $1.602 \times 10^{-19}$ | $3.827 \times 10^{-20}$ | $4.450 \times 10^{-26}$ | 1 | $10^{-6}$ | $1.783 \times 10^{-36}$ | $1.074 \times 10^{-9}$ |
| 1 million electron volts = | $1.519 \times 10^{-16}$ | $1.602 \times 10^{-6}$ | $1.182 \times 10^{-13}$ | $5.967 \times 10^{-20}$ | $1.602 \times 10^{-13}$ | $3.827 \times 10^{-14}$ | $4.450 \times 10^{-20}$ | $10^{6}$ | 1 | $1.783 \times 10^{-30}$ | $1.074 \times 10^{-3}$ |
| 1 kilogram = | $8.521 \times 10^{-13}$ | $8.987 \times 10^{23}$ | $6.629 \times 10^{16}$ | $3.348 \times 10^{10}$ | $8.987 \times 10^{16}$ | $2.147 \times 10^{16}$ | $2.497 \times 10^{10}$ | $5.610 \times 10^{35}$ | $5.610 \times 10^{29}$ | 1 | $6.025 \times 10^{26}$ |
| 1 atomic mass unit = | $1.415 \times 10^{-13}$ | $1.492 \times 10^{-3}$ | $1.100 \times 10^{-10}$ | $5.558 \times 10^{-17}$ | $1.492 \times 10^{-10}$ | $3.564 \times 10^{-11}$ | $4.145 \times 10^{-17}$ | $9.310 \times 10^{8}$ | 931.0 | $1.660 \times 10^{-27}$ | 1 |

1 meter – kgf = 9.807 joule, 1 watt-sec = 1 joule = 1 nt-meter, 1 cm-dyne = 1 erg

Some conversions used in spectroscopy:

1 eV = 8065.7 $cm^{-1}$ 1 $cm^{-1}$ = 0.000124 eV

1 eV ≈ 6000°K At 300°K, 3/2 kT ≈ 0.05 eV

**Notes :** The electron volt is the kinetic energy an electron gains from being accelerated through the potential difference of one volt in an electric field. The units enclosed by heavy lines are not properly energy units; they arise from the relativistic mass-energy equivalent formula $E = mc^2$.

## M. Specific Energy

| | $\frac{cal}{gm}$ | $\frac{erg}{gm}$ | $\frac{JOULE}{KGM}$ | $\frac{Btu}{lb_m}$ | $\frac{ft-lb_f}{lb_m}$ | $\frac{hp-hr}{lb_m}$ |
|---|---|---|---|---|---|---|
| 1 calorie pergram = | 1 | $4.186\times 10^{7}$ | $4.186\times 10^{3}$ | 1.800 | $1.400\times 10^{3}$ | $7.072\times 10^{-4}$ |
| 1 erg per gram = | $2.389\times 10^{-8}$ | 1 | $10^{-4}$ | $4.299\times 10^{-8}$ | $3.346\times 10^{-5}$ | $1.690\times 10^{-11}$ |
| 1 JOULE perKILOGRAM = | $2.389\times 10^{-4}$ | $10^{4}$ | 1 | $4.299\times 10^{-4}$ | 0.3346 | $1.690\times 10^{-7}$ |
| 1 Btu perpound (mass) = | 0.5557 | $2.326\times 10^{7}$ | $2.326\times 10^{3}$ | 1 | 777.9 | $3.929\times 10^{-4}$ |
| 1 foot-pound perpound (mass) = | $7.142\times 10^{-4}$ | $2.990\times 10^{4}$ | 2.990 | $1.285\times 10^{-3}$ | 1 | $5.051\times 10^{-7}$ |
| 1 horsepower-hourper pound (mass) = | $1.414\times 10^{3}$ | $5.920\times 10^{10}$ | $5.920\times 10^{6}$ | 2.545 | $1.980\times 10^{6}$ | 1 |

## N. Specific Energy per unit Temperature

| | $\frac{cal}{gm^{\circ}C}$ | $\frac{erg}{gm^{\circ}C}$ | $\frac{JOULE}{KGM^{\circ}K}$ | $\frac{Btu}{lb_m{}^{\circ}F}$ | $\frac{ft-lb_f}{lb_m{}^{\circ}F}$ | $\frac{hp-hr}{lb_m-^{\circ}F}$ |
|---|---|---|---|---|---|---|
| 1 calorie per gram per degree C = | 1 | $4.186\times 10^{7}$ | $4.186\times 10^{3}$ | 1.000 | 777.9 | $3.929\times 10^{-4}$ |
| 1 erg per gram per degree C = | $2.389\times 10^{-8}$ | 1 | $10^{-4}$ | $2.388\times 10^{-8}$ | $1.859\times 10^{-5}$ | $9.376\times 10^{-12}$ |
| 1 JOULE per KGM per DEGREE K = | $2.389\times 10^{-4}$ | $10^{4}$ | 1 | $2.388\times 10^{-4}$ | 0.1859 | $9.376\times 10^{-8}$ |
| 1 Btu per lb (mass) per degree F = | 1.000 | $4.187\times 10^{7}$ | $4.187\times 10^{3}$ | 1 | 777.9 | $3.929\times 10^{-4}$ |
| 1 foot-lb per lb (mass) per degree F = | $1.286\times 10^{-3}$ | $5.382\times 10^{4}$ | 5.382 | $1.285\times 10^{-3}$ | 1 | $5.051\times 10^{-7}$ |
| 1 horsepower-hour per lb (mass) per degree F = | $2.546\times 10^{3}$ | $1.066\times 10^{11}$ | $1.066\times 10^{7}$ | 2.545 | $1.980\times 10^{6}$ | 1 |

**Note (M & N):** The engineering units enclosed within the heavy lines have been properly related to the pound mass rather than the pound force because these specific thermal quantities depend on unit mass and have nothing to do with weight.

However, in engineering practice it is customary to relate energy and energy per degree to weight. Thus we speak of Btu/lb, ft-lb/lb and hp-hr/lb of weight. The conversion factors given in Tables M & N are equally valid for this purpose if the local acceleration of gravity if the earth standard value of g = 32.174 ft/sec$^2$ = 9.80665 meter/sec$^2$. This is true because the pound-force and the pound-mass are numerically equal at standard gravity. It should be realized that relating specific quantities to weight, rather than mass, involves a change of concept because weight and mass are not dimensional equivalents. The relation between units of mass and weight is not a relation between the concepts of mass and weight. The units are related by

$lb_f = 32.174\ lb_m\ ft/sec^2$

## O. Power

| | | $\frac{Btu}{hr}$ | $\frac{Btu}{sec}$ | $\frac{ft-lb}{min}$ | $\frac{ft-lb}{sec}$ | $\frac{cal}{sec}$ | $\frac{cal}{sec-cm^2}$ | **kw** | **WATT** |
|---|---|---|---|---|---|---|---|---|---|
| 1 British thermalunit per hour | = | 1 | $2.778\times 10^{-4}$ | 12.97 | 0.2161 | $3.929\times 10^{-4}$ | $7.000\times 10^{-2}$ | $2.930\times 10^{-4}$ | 0.2930 |
| 1 British thermal unit per second | = | 3600 | 1 | $4.669\times 10^{4}$ | 777.9 | 1.414 | 252.0 | 1.055 | $1.055\times 10^{3}$ |
| 1 foot-pound perminute | = | $7.713\times 10^{-2}$ | $2.142\times 10^{-5}$ | 1 | $1.667\times 10^{-2}$ | $3.030\times 10^{-5}$ | $5.399\times 10^{-3}$ | $2.260\times 10^{-5}$ | $2.260\times 10^{-2}$ |
| 1 foot-pound persecond | = | 4.628 | $1.286\times 10^{-3}$ | 60 | 1 | $1.818\times 10^{-3}$ | 0.3239 | $1.356\times 10^{-3}$ | 1.356 |
| 1 horsepower | = | 2545 | 0.7069 | $3.3\times 10^{4}$ | 550 | 1 | 178.2 | 0.7457 | 745.7 |
| 1 calorie persecond | = | 14.29 | 0.3950 | $1.852\times 10^{2}$ | 3.087 | $5.613\times 10^{-3}$ | 1 | $4.186\times 10^{-3}$ | 4.186 |
| 1 kilowatt | = | 3413 | 0.9481 | $4.425\times 10^{4}$ | 737.6 | 1.341 | 238.9 | 1 | 1000 |
| 1 WATT | = | 3.413 | $9.481\times 10^{-4}$ | 44.25 | 0.7376 | $1.341\times 10^{-3}$ | 0.2389 | 0.001 | 1 |

## P. Heat Flux*

| | | $\frac{cal}{sec-cm^2}$ | $\frac{kilocal}{hr-m^2}$ | $\frac{WATT}{M^2}$ | $\frac{watt}{in^2}$ | $\frac{Btu}{hr-ft^2}$ | $\frac{Btu}{sec-in^2}$ | $\frac{hp}{ft^2}$ |
|---|---|---|---|---|---|---|---|---|
| 1 calorie per sec per centimeter$^2$ | = | 1 | $3.600\times 10^{4}$ | $4.185\times 10^{4}$ | 27.00 | $1.327\times 10^{4}$ | $2.560\times 10^{-2}$ | 5.212 |
| 1 kilocalorie per hour per meter$^2$ | = | $2.778\times 10^{-5}$ | 1 | 1.163 | $7.500\times 10^{-4}$ | 0.3687 | $7.112\times 10^{-7}$ | $1.448\times 10^{-4}$ |
| 1 WATT per METER$^2$ | = | $2.390\times 10^{-5}$ | 0.8602 | 1 | $6.452\times 10^{-4}$ | 0.3171 | $6.117\times 10^{-7}$ | $1.246\times 10^{-4}$ |
| 1 watt per inch$^2$ | = | $3.704\times 10^{-2}$ | 1.333 | 1550 | 1 | 491.5 | $9.481\times 10^{-4}$ | 0.1931 |
| 1 British thermal unit per hour per foot$^2$ | = | $7.535\times 10^{-5}$ | 2.713 | 3.153 | $2.035\times 10^{-3}$ | 1 | $1.929\times 10^{-6}$ | $3.928\times 10^{-4}$ |
| 1 British thermal unit per sec per inch$^2$ | = | 39.06 | $1.406\times 10^{6}$ | $1.635\times 10^{6}$ | $1.055\times 10^{3}$ | $5.184\times 10^{5}$ | 1 | 203.6 |
| 1 horsepower per foot$^2$ | = | 0.1918 | $6.905\times 10^{2}$ | 8027 | 5.179 | $2.546\times 10^{3}$ | $4.911\times 10^{-3}$ | 1 |

*Also power per unit area

## Q. Heat Transfer Coefficient, h

| | | $\frac{cal}{sec-_{cm}^{2}\ ^{\circ}C}$ | $\frac{WATT}{M^{2^\circ}K}$ | $\frac{watt}{in^2\ ^\circ C}$ | $\frac{Btu}{hr-ft^{2\circ}F}$ | $\frac{Btu}{sec-in^{2\circ}F}$ | $\frac{hp}{tt^{2\circ}F}$ |
|---|---|---|---|---|---|---|---|
| 1 calorie per secper centimeter$^2$ - ºC | = | 1 | $4.185\times 10^4$ | 27.00 | $7.372\times 10^3$ | $1.422\times 10^{-2}$ | 2.895 |
| 1 WATT per METER$^2$per DEG KELVIN | = | $2.390\times 10^{-5}$ | 1 | $6.452\times 10^{-4}$ | 0.1762 | $3.398\times 10^{-7}$ | $6.922\times 10^{-5}$ |
| 1 watt per inch$^2$ perdeg Centigrade | = | $3.704\times 10^{-2}$ | 1550 | 1 | 273.1 | $5.267\times 10^{-4}$ | 0.1073 |
| 1 Btu per hour perper foot$^2$ - ºF | = | $1.356\times 10^{-4}$ | 5.675 | $3.663\times 10^{-3}$ | 1 | $1.929\times 10^{-6}$ | $3.928\times 10^{-4}$ |
| 1 Btu per sec perinch$^2$ - ºF | = | 70.31 | $2.943\times 10^6$ | $1.899\times 10^3$ | $5.184\times 10^5$ | 1 | 203.6 |
| 1 horsepower perfoot$^2$ - ºF | = | 0.3452 | $1.445\times 10^4$ | 9.322 | $2.546\times 10^3$ | $4.911\times 10^{-3}$ | 1 |

## R. R. Thermal Conductivity, k

| | | $\frac{cal}{sec-cm^\circ C}$ | $\frac{WATTS}{METER^\circ K}$ | $\frac{watts}{in^\circ C}$ | $\frac{Btu}{hr-ft-^\circ F}$ | $\frac{Btu}{sec-in^\circ F}$ | $\frac{hp}{ft^\circ F}$ |
|---|---|---|---|---|---|---|---|
| 1 calorie per sec per centimeter-deg C | = | 1 | 418.5 | 10.63 | 241.9 | $5.600\times 10^{-3}$ | $9.503\times 10^{-2}$ |
| 1 WATT per METERper DEG KELVIN | = | $2.390\times 10^{-3}$ | 1 | $2.540\times 10^{-2}$ | 0.5781 | $1.338\times 10^{-5}$ | $2.271\times 10^{-4}$ |
| 1 watt per inch perdeg Centigrade | = | $9.407\times 10^{-2}$ | 39.37 | 1 | 22.76 | $5.269\times 10^{-4}$ | $8.939\times 10^{-3}$ |
| 1 Btu per hour perfoot-deg F | = | $4.134\times 10^{-3}$ | 1.730 | $4.394\times 10^{-2}$ | 1 | $2.315\times 10^{-3}$ | $3.929\times 10^{-4}$ |
| 1 Btu per sec perinch-deg F | = | $1.786\times 10^2$ | $7.474\times 10^4$ | $1.898\times 10^3$ | $4.320\times 10^4$ | 1 | 16.97 |
| 1 horsepower perfoot-deg F | = | 10.52 | 4403 | 111.8 | 2546 | $5.894\times 10^{-2}$ | 1 |

## S. Absolute or Dynamic Viscosity, μ

| | centipoise | poise | $\frac{kgm_f - sec}{meter^2}$ | $\frac{lb - esc}{ft^2}$ | $\frac{KGM}{M - SEC}$ | $\frac{lb_m}{ft - sec}$ |
|---|---|---|---|---|---|---|
| 1 centipoise = | 1 | $10^{-2}$ | $1.020 \times 10^{-4}$ | $2.089 \times 10^{-5}$ | $10^{-3}$ | $6.720 \times 10^{-4}$ |
| 1 poise = | 100 | 1 | $1.020 \times 10^{-2}$ | $2.089 \times 10^{-3}$ | 0.100 | $6.720 \times 10^{-2}$ |
| 1 kg (force) – sec per meter² = | $9.807 \times 10^3$ | 98.07 | 1 | 0.2048 | 9.807 | 6.590 |
| 1 lb (force) – sec per foot ² = | $4.788 \times 10^4$ | $4.788 \times 10^2$ | 4.882 | 1 | 47.88 | 32.174 |
| 1 KILOGRAM per METER-SEC = | $10^3$ | 10 | 0.1020 | $2.089 \times 10^{-2}$ | 1 | 0.6720 |
| 1 lb (mass) per foot – sec = | $1.488 \times 10^3$ | 14.88 | 0.1518 | $3.108 \times 10^{-2}$ | 1.488 | 1 |

**Note** : The absolute viscosity μ is properly expressed in force units according to its definition. In heat transfer and fluid mechanics it is usually expressed in mass- equivalent units to avoid the use of a conversion factor in Reynolds Number. Mass units have been used in the portion of the table enclosed in heavy lines. The proper force units for μ in the mksq system are NEWTON-SEC per METER²; they are seldom used. The poise is the cgs force unit and is defined by

$$1\,\text{poise} = 1\frac{\text{dyne}}{\text{second centimeter}^2}$$

## T. Kinematic Viscosity, ν = μ/ρ

| | | **centistoke** | **stoke** | **METER²/SEC** | **ft²/sec** |
|---|---|---|---|---|---|
| 1 centistoke | = | 1 | $10^{-2}$ | $10^{-6}$ | $1.076 \times 10^{-5}$ |
| 1 stoke | = | 100 | 1 | $10^{-4}$ | $1.076 \times 10^{-3}$ |
| 1 METER²/SEC | = | $10^6$ | $10^4$ | 1 | 10.76 |
| 1 ft²/sec | = | $9.290 \times 10^4$ | 929.0 | $9.290 \times 10^{-2}$ | 1 |

1 stoke = 1 centimeter²/sec

## AA. Electric Charge

| | | abcoul | amp-hr | COUL | faraday | statcoul |
|---|---|---|---|---|---|---|
| 1 abcoulomb (1 emu) | = | 1 | $2.778 \times 10^{-3}$ | 10 | $1.036 \times 10^{-4}$ | $2.998 \times 10^{10}$ |
| 1 ampere-hour | = | 360 | 1 | 3600 | $3.730 \times 10^{-2}$ | $1.079 \times 10^{13}$ |
| 1 COULOMB | = | 0.100 | $2.778 \times 10^{-4}$ | 1 | $1.036 \times 10^{-5}$ | $2.998 \times 10^9$ |

| | | | | | | |
|---|---|---|---|---|---|---|
| 1 faraday | = | $9.652 \times 10^{3}$ | 26.81 | $9.652 \times 10^{4}$ | 1 | $2.893 \times 10^{14}$ |
| 1 statcoulomb (1 esu) | = | $3.336 \times 10^{-11}$ | $9.266 \times 10^{-14}$ | $3.336 \times 10^{-10}$ | $3.456 \times 10^{-15}$ | 1 |

1 electronic charge $= 1.602 \times 10^{-19}$ coulombs

$= (1.602 \times 10^{-19})\,(2.998 \times 10^{9})$ statcoulomb

$= 4.8 \times 10^{-10}$ esu

## BB. Electric Current

| | | abamp | AMP | statamp |
|---|---|---|---|---|
| 1 abampere (1 emu) | = | 1 | 10 | $2.998 \times 10^{10}$ |
| 1 AMPERE | = | 0.100 | 1 | $2.998 \times 10^{9}$ |
| 1 statampere (1 esu) | = | $3.336 \times 10^{-11}$ | $3.336 \times 10^{-10}$ | 1 |

## CC. Electric Potential, Electromotive Force

| | | abv | VOLT | statv |
|---|---|---|---|---|
| 1 abvolt (1 emu) | = | 1 | $10^{-8}$ | $3.336 \times 10^{-11}$ |
| 1 VOLT | = | $10^{8}$ | 1 | $3.336 \times 10^{-3}$ |
| 1 statvolt (1 esu) | = | $2.998 \times 10^{10}$ | 299.8 | 1 |

## DD. Electric Resistance

| | | abohm | OHM | statohm |
|---|---|---|---|---|
| 1 abohm (1 emu) | = | 1 | $10^{-9}$ | $1.113 \times 10^{-21}$ |
| 1 OHM | = | $10^{9}$ | 1 | $1.113 \times 10^{-12}$ |
| 1 statohm (1 esu) | = | $8.987 \times 10^{20}$ | $8.987 \times 10^{11}$ | 1 |

## EE. Electric Resistivity, Reciprocal Conductivity

| | | abohm-cm | ohm-cm | OHM-M | statohm-cm | ohm-circ mil/ft |
|---|---|---|---|---|---|---|
| 1 abohm-centimeter (1 emu) | = | 1 | $10^{-9}$ | $10^{-11}$ | $1.113 \times 10^{-21}$ | $6.015 \times 10^{-3}$ |
| 1 ohm-centimeter | = | $10^{9}$ | 1 | 0.0100 | $1.113 \times 10^{-12}$ | $6.015 \times 10^{6}$ |
| 1 OHM-METER | = | $10^{11}$ | 100 | 1 | $1.113 \times 10^{-10}$ | $6.015 \times 10^{8}$ |

| | | | | | |
|---|---|---|---|---|---|
| 1 statohm-centimeter (1 esu) = | 8.987 $\times 10^{20}$ | 8.987 $\times 10^{11}$ | 8.987 $\times 10^{9}$ | 1 | 5.406 $\times 10^{18}$ |
| 1 ohm-circular mil per foot = | 166.2 | 1.662 $\times 10^{-7}$ | 1.662 $\times 10^{-9}$ | 1.850 $\times 10^{-19}$ | 1 |

## FF. Capacitance

| | abf | FARAD | μf | statf |
|---|---|---|---|---|
| 1 abfarad (1 emu) = | 1 | $10^{9}$ | $10^{15}$ | $8.987 \times 10^{20}$ |
| 1 FARAD = | $10^{-9}$ | 1 | $10^{6}$ | $8.987 \times 10^{11}$ |
| 1 microfarad = | $10^{-15}$ | $10^{-6}$ | 1 | $8.987 \times 10^{5}$ |
| 1 statfarad (1 esu) = | $1.113 \times 10^{-21}$ | $1.113 \times 10^{-12}$ | $1.113 \times 10^{-6}$ | 1 |

## GG. Inductance

| | | abhenry | HENRY | μh | stathenry |
|---|---|---|---|---|---|
| 1 abhenry (1 emu) | = | 1 | $10^{-9}$ | 0.001 | $1.113 \times 10^{-21}$ |
| 1 HENRY | = | $10^{9}$ | 1 | $10^{6}$ | $1.113 \times 10^{-12}$ |
| 1 microhenry | = | $10^{3}$ | $10^{-6}$ | 1 | $1.113 \times 10^{-18}$ |
| 1 stathenry (1 esu) | = | $8.987 \times 10^{20}$ | $8.987 \times 10^{11}$ | $8.987 \times 10^{17}$ | 1 |

## HH. Magnetic Flux

| | maxwell | kiloline | WEBER |
|---|---|---|---|
| 1 maxwell (1 line or 1 emu) = | 1 | 0.001 | $10^{-8}$ |
| 1 kiloline = | 1000 | 1 | $10^{-5}$ |
| 1 WEBER = | $10^{8}$ | $10^{5}$ | 1 |

1 esu = 2.998 webers

## II. Magnetomotive Force

| | abamp-turn | AMP-TURN | gilbert |
|---|---|---|---|
| 1 abamp-turn = | 1 | 10 | 12.57 |
| 1 AMP-TURN = | 0.100 | 1 | 1.257 |
| 1 gilbert = | 7.958 $\times 10^{-2}$ | 0.7958 | 1 |

1 pragilbert = 4 $\pi$ amp-turn

1 esu = $2.655 \times 10^{-11}$ amp-turn

## JJ. Magnetic Field Strength, B

| | gauss | $\frac{\text{kiloline}}{\text{in}^2}$ | $\frac{\text{WEBER}}{\text{METER}^2}$ | milligauss | gamma |
|---|---|---|---|---|---|
| 1 gauss (line per $cm^2$) = | 1 | $6.452 \times 10^{-2}$ | $10^{-4}$ | 1000 | $10^5$ |
| 1 kiloline per $in^2$ = | 155.0 | 1 | $1.550 \times 10^{-2}$ | $1.550 \times 10^5$ | $1.550 \times 10^7$ |
| 1 WEBER PER $METER^2$ = | $10^4$ | 64.52 | 1 | $10^7$ | $10^9$ |
| 1 milligauss = | $10^{-3}$ | $6.452 \times 10^{-6}$ | $10^{-7}$ | 1 | 100 |
| 1 gamma = | $10^{-5}$ | $6.452 \times 10^{-8}$ | $10^{-9}$ | $10^{-2}$ | 1 |

1 esu = $2.998 \times 10^6$ weber/meter$^2$

$10^4$ gauss = 1 tesla

## KK. Magnetic Vector, H

| | $\frac{\text{abamp – turn}}{\text{cm}}$ | $\frac{\text{amp – turn}}{\text{cm}}$ | $\frac{\text{AMP – TURN}}{\text{METER}}$ | $\frac{\text{amp – turn}}{\text{in}}$ | oersted |
|---|---|---|---|---|---|
| 1 abampere-turn per centimeter = | 1 | 10 | 1000 | 25.40 | 12.57 |
| 1 ampere-turn per centimeter = | 0.100 | 1 | 100 | 2.54 | 1.257 |
| 1 AMPERE-TURN PER METER = | $10^{-3}$ | $10^{-2}$ | 1 | $2.540 \times 10^{-2}$ | $1.257 \times 10^{-2}$ |
| 1 ampere-turn per inch = | $3.937 \times 10^{-2}$ | 0.3937 | 39.37 | 1 | 0.4947 |
| 1 oersted = | $7.958 \times 10^{-2}$ | 0.7958 | 79.58 | 2.021 | 1 |

1 oersted = 1 gilbert/cm

1 esu = $2.655 \times 10^{-9}$ amp-turn/meter

1 praoersted = 4 $\pi$ amp-turn/meter

## VIII. Electromagnetic Constants of Free Space

Maxwell was able to show analytically that the constant appearing in a wave equation derived for free space (perfect vacuum), for the case $\rho e = \sigma c = 0$, was the square of the velocity of propagation of electromagnetic waves in free space. The experiments of Hertz verified that this velocity was the velocity of light in free space and that

$$c^{1} = \frac{1}{\mu_0 \varepsilon_0}$$

*where* $\mu_0$ is the permeability of free space and $\varepsilon_0$ is the permittivity of free space. This equation is true for any system of units; in the mksq system

$c = 2.997925 \times 10^{8}$ meter/sec

$\mu_0 = 1.256637 \times 10^{-6}$ henry/meter

$\varepsilon_0 = 8.85416 \times 10^{-12}$ farad/meter

## IX. Electromagnetic Constants of Materials

The permeability and permittivity of materials are usually given relative to the values of free space.

Relative permeability

$$k_m = \frac{\mu}{0}$$

Magnetic Susceptibility

$$k_m = 1 + \chi_m$$

Relative permittivity

$$k_e = \frac{\varepsilon}{0}$$

(Dielectric constant)

Electric Susceptibility

$$\kappa e = 1 + \chi e$$

When looking up values of electromagnetic constants of materials care must be taken to be sure what values are specified, i.e. $\mu$, $\kappa_m$ or $\chi_m$ and $\varepsilon$, $\kappa_e$ or $\chi_e$. The usual values given are $\kappa_e$ and $\chi_m$.

$\kappa_m$ is a number near unity and may be greater or less than unity; $\chi_m$ may be positive or negative but is small compared to unity. $\kappa_e$ is always greater than unity and may be significantly larger than unity; $\chi_e$ is always positive and may be large compared to unity:

$\kappa_m \lesseqgtr 1, \kappa_m \sim 1$

$\chi_m \lesseqgtr 0, \chi_m << 1$

$\kappa_m$ is temperature dependent

$\kappa_c \geq 1$

$\chi_e \geq 0$

possible $\chi_e >> 1$

## X. Some Important Dimensional Constants (*mksq* units)

| Name | Symbol | Computational Value | Best Experimental Value |
|---|---|---|---|
| Speed of light | c | $3.00 \times 10^8$ meter/sec | $(2.997930 \pm 0.000003) \times 10^8$ |
| Universal Gravitational Const. | G | $6.67 \times 10^{-11}$ nt-m²/kgm² | $(6.673 \pm 0.003) \times 10^{-11}$ |
| Avogadro's Number | $N_0$ | $6.02 \times 10^{23}$ /mole | $(6.02486 \pm 0.00016) \times 10^{23}$ |
| Universal Gas Constant | R | 8.32 joule/mole °K | $8.31696 \pm 0.00034$ |
| Standard Volume of Ideal Gas | | $2.24 \times 10^{-2}$ meter³ | $(2.24207 \pm 0.00006) \times 10^{-2}$ |
| Planck's Constant | h | $6.63 \times 10^{-34}$ joule/sec | $(6.62517 \pm 0.00023) \times 10^{-34}$ |
| Boltzmann's Constant | k | $1.38 \times 10^{-23}$ joule/°K | $(1.38044 \pm 0.00007) \times 10^{-23}$ |
| Mechanical Equiv. of Heat | J | 4.19 joule/cal | $4.1855 \pm 0.0004$ |
| Triple Point of Water | | 273.16 °K | 273.16 °K exactly |
| Ice Point of Water | $\Sigma_0$ | 273.16 °K | 273.16 °K $\pm 0.0002$ |
| Maximum Density of Water (at 3.98°C, 1 atm) | | 1 gm/cm³ | 0.999973 |
| Permeability of Free Space | $\mu_0$ | $1.26 \times 10^{-6}$ henry/meter | $4\pi \times 10^{-7}$ exactly |
| Permittivity of Free Space | $\varepsilon_0$ | $8.85 \times 10^{-12}$ farad/meter | $(8.85415 \pm 0.00002) \times 10^{-12}$ |
| Electronic Charge | e | $1.60 \times 10^{-19}$ coulomb | $(1.60206 \pm 0.00003) \times 10^{-19}$ |
| Electronic Rest Mass | $m_e$ | $9.11 \times 10^{-31}$ kgm | $(9.1083 \pm 0.0003) \times 10^{-31}$ |
| Proton Rest Mass | $m_p$ | $1.67 \times 10^{-27}$ kgm | $(1.67239 \pm 0.00004) \times 10^{-27}$ |
| Neutron Rest Mass | $m_n$ | $1.67 \times 10^{-27}$ kgm | $(1.67470 \pm 0.00004) \times 10^{-27}$ |
| Mass-Energy Relation | $c^2 = E/m$ | $8.99 \times 10^{16}$ meter²/sec² | $(8.98758 \pm 0.00003) \times 10^{16}$ |
| Magnetic Moment of Electron | | $9.28 \times 10^{-32}$ joule-m²/weber | $(9.2837 \pm 0.0002) \times 10^{-32}$ |
| Compton Wavelength of Electron | $\lambda_c$ | $2.43 \times 10^{-12}$ meter | $(2.42626 \pm 0.00002) \times 10^{-12}$ |
| First Bohr Orbit Radius in | $a_0$ | $5.29 \times 10^{-11}$ meter | $(5.29172 \pm 0.00002) \times 10^{-11}$ |
| Hydrogen Atom | | | |
| Stefan-Boltzmann Const. | σ | $5.67 \times 10^{-8}$ joule/sec (°K)⁴ meter² | $(5.6687 \pm 0.0019) \times 10^{-8}$ |

## XI. Some Important Dimensionless Groups

| Name | Group | Field of Use |
|---|---|---|
| Biot | (Bi) = $hL/k$ | Conduction heat transfer |
| Euler | (Eu) = $p/\rho V^2$ | Fluid mechanics |
| Fourier | (Fo) = $\alpha\tau/L^2$ | Conduction heat transfer |
| Froude | (Fr) = $V\sqrt{Lg}$ | Fluid mechanics |
| Graetz | (Grz) = $w$ & $c_p/kL$ | Heat transfer, free convection Grashof |
| Grashof | (Grf) = $L^3\rho^2 g\beta\Delta t/\mu^2$ | Heat transfer, free convection |
| Hartmann | (Ha) = ($\sigma_c$ $B$ 2 $L$2 / μ)1 / 2 | MHD |
| Knudsen | (Kn) = $\lambda/L$ | Fluid mech., rarified gas flow Lewis |
| Lewis | (Le) = (Sc)/(Pr) = $\alpha/D$ | Conv. heat & mass transfer Mach |
| Mach | (M) = $V/a$ | High speed flow |
| Nusselt | (Nu) = $hL/k$ | Convection heat transfer |
| Péclet | (Pe) = (Re)(Pr) | Convection heat transfer |
| Prandtl | (Pr) = $\mu c_p/k$ | Convection heat transfer |
| Reynolds | (Re) = $VL\rho/\mu$ | Fluid mech., heat transfer |
| Magnetic Reynolds | $(Re)_m = \mu\sigma_c VL$ | MHD |
| Schmidt | (Sc) = $\mu/\rho D$ | Conv. heat & mass transfer |
| Stanton | (St) = $h/c_p V\rho$ | Convection heat transfer |
| Weber | (We) = $\rho V^2 L/\sigma$ | Fluid mechanics, free surface |
| | (N) = $(Ha)^2/(Re)$ | MHD |
| | (S) = $(Ha)^2/(Re)(Re)_m$ | MHD |

Symbols

$B_0$ – Applied mag. field

$D$ – Diffusion coefficient

$L$ – Characteristic length

$V$ – Fluid velocity

$a$ – Acoustic velocity

$g$ – Gravity

$h$ – Heat transfer coefficient

$k$ – Thermal conductivity

$p$ – Static pressure

ρ – Density

λ - Mean free path (molecular)

$\Delta t$ – Temperature difference

$w$& - Mass rate of flow

$c_p$ – Specific heat (const. pressure)

α - Thermal diffusivity

β - Thermal expansion coefficient

σ - Surface tension

$\sigma_c$ – Electric conductivity

μ - Viscosity or magnetic permeability

τ - Time interval

## XII. The Perfect Gas Law

### A. Nomenclature, Definitions and Equations

In the following discussion the elementary particle under consideration is the molecule. Care must be taken to use consistent units, especially to make the proper distinction between mass and weight units. The units of the gas constant must be consistent with those used for density or specific volume.

### Nomenclature

| **Symbol** | **Definition in cgs Units** |
|---|---|
| $\rho$ | Density in grams per $cm^3$ |
| $k$ | Boltzmann's Constant in ergs/molecule K° |
| **m** | Mass of molecule in grams |
| **n** | Particle Number Density in molecules/$cm^3$ |
| $p$ | Pressure in dynes/ $cm^2$ |
| $\upsilon$ | Specific Volume in $cm^3$ per gram |
| $\upsilon_N$ | Specific Molar Volume in $cm^3$ /gm-mole |
| **M** | Molecular Weight in gm/gm-mole |
| **N** | Number of moles |
| $N_0$ | Avogadro's Number, no. of molecules per gm-mole |
| $R$ | Gas Constant in ergs/gm K° |
| $R_0$ | Universal Gas Constant in ergs/gm-mole K° |
| $T$ | Absolute Temperature, degrees Kelvin |
| **V** | Total Volume in cubic centimeters |
| **W** | "Weight" of Gas in grams |

Avogadro's Law: All ideal gases at the same temperature and pressure have the same specific molar volume. At STP (0°C, 1 atm), $\upsilon_N$ = 22.4 liters/gm-mole = 22400 $cm^3$/gm- mole = 359 $ft^3$/lb-mole. At the same temperature and pressure all ideal gases contain the same number of molecules per unit volume.

Definition of the mole: The gm-mole is the amount of an ideal gas which will occupy the same volume as 32 gm of oxygen at STP; the pound-mole is similarly related to 32 lb of oxygen. The weight of a mole of gas is numerically equal to the molecular weight of the gas.

## B. Values of Universal Gas Constant, Boltzmann's Constant and Avogadro's Number in Different Units

| Units | $R_0$ | Units | $R_0$ |
|---|---|---|---|
| ft-lb/lb-mole R° | 1544 | cal/gm-mole K° | 1.987 |
| ft-lb/lb-mole K° | 2779 | $cm^3$ atm/gm-mole K° | 82.06 |
| ft-lb/lb-mole R° | 3.407 | liter atm/gm-mole K° | 0.08206 |
| Btu/lb-mole R° | 1.987 | ergs/gm-mole K° | $8.313 \times 10^7$ |
| $ft^3$ atm/lb-mole R° | 0.729 | JOULES/KGM-MOLE K° | 8313 |
| $ft^3$ atm/lb-mole K° | 1.315 | psia $ft^3$/lb-mole R° | 10.71 |

| Units | k | Units | $N_0$ |
|---|---|---|---|
| ergs/molecule K° | $1.38 \times 10^{-16}$ | molecules/gm-mole | $6.023 \times 10^{23}$ |
| JOULES/ MOLECULE K° | $1.38 \times 10^{-23}$ | MOLECULES/ KGM-MOLE | $6.023 \times 10^{26}$ |
| ft-lb/molecule R° | $5.655 \times 10^{-24}$ | molecules/lb-mole | $2.73 \times 10^{26}$ |
| Btu/molecule R° | $7.267 \times 10^{-27}$ | | |
| 1 atm = $1.013 \times 10^6$ dynes/$cm^2$ | | 1 erg = $10^{-7}$ joule | |
| 1 erg = 1 dyne-cm | | 1 ft-lb = 1.356 $10^7$ ergs | |

# Bibliography


A.M.Takale & V.R. Naik, "Design & manufacturing of multi spindle drilling head (msdh) for its cycle time optimization", International Journal of Mechanical Engineering applications Research – IJMEAR, Vol 03, Issue 01; January-April 2012.

A.S.Udgave,Prof.V.J.Khot, Design & development of multi spindle drilling head (msdh), IOSR Journal of Mechanical and Civil Engineering (IOSR-JMCE) ISSN: 2278-1684, PP: 60-69

Afolalu S.A., Samuel O.D., Ikumapayi M.O., Oladipupo S., Emetere M. (2020). Study of Mechanical Behavious and Characterization of a Steel Joint in MIG Welding Under Varying Fluxes. International Journal of Engineering Research and Technology, 13, 6, 1300-1306.

ANSI B11.8 Drilling, Milling, and Boring Machines. https://webstore.ansi.org standards/amt/ansib112021

Anupam Chaturvedi, Prof. Jyoti Mishra, Prof. Vijay Parmar, "An Improved Scissor Lift working on Lead Screw Mechanism Aerial Scissor Lift and its Accessories", Article in International Journal of Advance Engineering and Research, March 2017.

Arora, R. P. Basic Mechanical Engineering. Charotar Publishing House Pvt. Ltd., Anand (Guj.).

Balkeshwar Singh, Anil Kumar Mishra "Analysis and Fabrication of Remote Control Lifting Jack", Article in International Journal of Scientific Engineering and Applied Science, June 2015.

Banjo, S.O. Bolaji, B.O., Ajayi, O.O., Olufemi, B.P., Osagie, I. and Onokwai, A.O. (2019). Performance enhancement using appropriate mass charge of R600a in a developed domestic refrigerator, IOP Conf. Ser.: Earth and Environmental Science, 331 (1), 1-8.

Banjo, S.O. Bolaji, B.O., I. Osagie, O.S.I. Fayomi, O.B. Fakehinde, P.S. Olayiwola, S.O. Oyedepo, N.E. Udoye (2019). Experimental Analysis of the performance characteristics of an eco-friendly HC600a as a retrofitting refrigerant in a thermal system, Journal of Physics: Conf. Ser. 1378 (4), 1-8.

Bhavin.H. Mehta, Mehul.G. Joshi, Utsav.M.Dholakiya, "Design and Analysis of Screw Operated Lifting Equipment", Article in International Journal for Scientific Research and Development, 2016.

Black J.T and Kohser R.A. (2012). Materials and Processes in Manufacturing, ($11^{th}$ Ed.), John Wiley and Sons, Inc. Printed in the United States of America.

Chapman, W.A.J. and Martin, S. J. 1998. Workshop Technology. Part-III. Viva Books Private Ltd, New Delhi.

Choudhary, H. S. K. (2010). Elements of Workshop Technology Vol. 1. Media Promoters & Publishers Pvt. Ltd., New Delhi.

Choudhary, H. S. K. (2010). Elements of Workshop Technology Vol. 2. Media Promoters & Publishers Pvt. Ltd., New Delhi.

Divyesh.Prafulla.Ubale, Alan Francy, N.P Sherje, "Design, Analysis and Development of Multiutility home equipment using Scissor Lift Mechanism", Article in International Journal of Scientific Research and Management, 2015

Gupta, H.N., Gupta, R.C. and Mittal, A. (2009). Manufacturing Processes, (2nd Ed.), New Age International (P) Limited Publishers.
Hodder and Stoughton (1977). Basic Mech. Engineering. Printed in Great Britain by Westminster press group.
http://uhv.cheme.cmu.edu/procedures/machining/ch4.pdf
http://www.ignou.ac.in/upload/bme059unit-1.pdf
http://www.jnkvv.org/PDF/07052020101119652017 48.pdf
https://astromachineworks.com/what-is-cnc-machining/
https://books.google.co.in/books?id=ijIfEAAAQBAJ&pg=RA2-PA29&lpg=RA2&redir_esc=y#v=onepage&q&f=false
https://cncmachines.com/what-is-a-cnc-machine
https://dl-manual.com/doc/power-electronics-lab-manual-go3d960m25v8
https://en.wikipedia.org/wiki/Milling (machining)
https://esskaymachines.com/blog/%EF%BB%BFshaper-machine/
https://gptcadoor.org/assets/downloads/f1hmtu025lsby13.pdf
https://iopscience.iop.org/article/10.1088/1757-899X/1107/1/012194/pdf
https://jyoti.co.in/
https://learnmech.com/shaper-machine/
https://learnmechanical.com/planer-machine/
https://mechanicalenotes.com/fitting-workshop/
https://mechanicalenotes.com/shaper-machine/
https://mechanicalnotes.com/lathe-machine-introduction/
https://mechanicalnotes.com/shaper-machine/
https://phillipscorp.com/india/milling-machine-process-parts-types-uses
https://themechanicalengineering.com/
https://themechanicalengineering.com/drilling-machine/
https://themechanicalengineering.com/lathe-machine/
https://themechanicalengineering.com/milling-machine/
https://themechanicalengineering.com/shaper-machine/
https://www.bhavyamachinetools.com/all-geared-lathe-machine
https://www.britannica.com/
https://www.britannica.com/technology/milling-machine
https://www.cncmasters.com/cnc-machine-buyers-guide/
https://www.cncmasters.com/lathe-machine-buyers-guide/
https://www.coursehero.com/file/14317807/5-20-Metalwork-Operations-Sheet-Metal-Work/
https://www.ehs.harvard.edu/sites/default/files/machine_shop_safety_osha_guidance_milling_machine.pdf
https://www.engineeringarticles.org/milling-machine-definition-process-types/
https://www.engineersgallery.com/general-press-operations/
https://www.irjet.net/archives/V8/i12/IRJET-V8I12251.pdf
https://www.mechtechguru.com/2020/05/principle-parts-of-column-andkneetype. html
https://www.researchgate.net/
https://www.researchgate.net/publication/348355557_Fabrication_of_Universal_Drilling_Machine
https://www.researchgate.net/publication/362520556_Design_and_Manufacturing_of_Portable_Drilling_Machine
https://www.sathyabama.ac.in/sites/default/files/course-material/2020-10/UNIT-2_12.pdf
https://www.studocu.com/in/document/delhi-technological-university/mechanical/study-of-drilling-machine/23301993
https://www.techtarget.com/searcherp/definition/computer-numerical-control-CNC

https://www.thecrucible.org/guides/machining/lathes/

https://www.theengineerspost.com/15-different-types-of-milling-machines/

https://www.theengineerspost.com/types-of-shaper-machines/

https://www.thomasnet.com/articles/custom-manufacturing-fabricating/understanding-cnc-machining/

Images: Google.

Ivan Sunit Rout, Dipti Ranjan Patra, Siddhartha Sankar Padhi, "Design and Fabrication of motorized automated Object lifting jack", Article in IOSR Journal of Engineering, May 2014.

Jay M. PatelMr. Akhil P. NairProf. Hiral U.Chauhan, "3-Directional Flexible Drilling Machine", IJSRD - International Journal for Scientific Research & Development| Vol. 3, Issue 01, 2015 | ISSN (online): 2321-0613, PP: 1262 to 1264.

Kashyap, S.K., Kumar, S., Mallick, M., Singh, R.P., Verma, M. (2018). A comparative study between experimental and theoretical buckling load for hollow steel column. International Journal of Engineering, Science and Technology, 10 (3), 27-33.

Kdenigsberger, F. and Tobias, S.A. (1973). Machine Tools Design and Research Proceeding of the Fourteen International Conference. The Macmillan Press Limited, London, pg.125-132.

Kiran, R. P. (2015). Workshop Technology, Vikas Publishing House Pvt Limited.

Kumar, J.S.S. and Sekhar, S.R.S. (2015). An Experiment Study on the Influence Cutting Parameters on Thrust Force" While Drilling on Different Materials. International Journal and Magazine of Engineering Technology, Management and Research, 2 (1) 327-336.

Lab Manual of IIT Kanpur. https://home.iitk.ac.in/~vkjain/manual.pdf

Malode, T.B. and Gandhe, R.R. (2016). Designed and fabrication of multi-spindle machine. International Journal of Innovative Research in Science and Technology, 3 (2), 290-295.

O. P. Khanna's Production Technology Manufacturing Processes Volume-I, Dhanpat Rai Publications, New Delhi.

Oregon OSHA Division 2/Subdivision O 29 CFR 1910.212 — General Requirements for All Machines https://osha.oregon.gov/OSHAPubs/2980.pdf

Parashar, B.S.N and Mittal, S.K. (2011). Elements of Manufacturing Processes, 10th edition, PHI Learning Private Limited, 259-363.

Prakash R.S., Singh, R.D., Singh, B., Prakash, S.A. and Koshy, V. (2018). Design and Fabrication of Multi Axis Drilling Machine, International Journal of Innovative Research in Technology, 4(11), 702-706.

R. A. Tatara, "Design and Construction of a Compact Portable All Electric Scissor Lift", Article in Academic Journal of Manufacturing Engineering, Jan 2018.

Raghuvanshi, B. S. 2002. Workshop Technology. Vol. 2. Dhanpat Rai & Co. (P) Ltd, Delhi.

Shinde V., Panchal R.N., Panchal J.R. (2016). Semi-automatic Drilling Machine Controller. International Journal of Innovative Research in Science, Engineering and Technology, 5 (5), 7522-7529.

Singh A.K. and Moona M. (2016). Design of a Universal Micro Radial Drilling Machine. International Journal of Science Technology and Engineering, 2 (9), 154-163.

Widiyono E., Winarto, Wardhani R., Rusdiyana L. (2011). Engineering Design of a Gang Drilling Machine Equipped with Jig and Fixtures to make A Prototype Machine in Birdcage Production, The Journal for Technology and Science, 22, 4, 214-220.

Workshop manual of Sri Venkateswara College of Engineering, Chittoor.

Workshop practise Laboratory manual of Dayananda Sagar College of Engineering.

Xavier L.F and Elangova D. (2013). Effective Parameters for Improving Deep Hole Drilling Process by Conventional Method. International Journal of Engineering Research and Technology, 2 (3), 1-11.

# Index